Churchill's Spearhead

To Jo-anne and Alice

Churchill's Spearhead

The Development of Britain's Airborne Forces during the Second World War

John William Greenacre PhD

Pen & Sword
AVIATION

First published in
Great Britain in 2010
By Pen and Sword Aviation
An imprint of
Pen and Sword Books Ltd
47 Church Street
Barnsley
South Yorkshire
S70 2AS

Copyright © John William Greenacre 2010

ISBN 978-1-84884-271-7

The right of John William Greenacre to be identified as the Author of this Work has been asserted by him in accordance with the Copyright, Designs and Patents Act 1988.

A CIP record for this book is available from the British Library

All rights reserved. No part of this book may be reproduced or transmitted in any form or by any means, electronic or mechanical including photocopying, recording or by any information storage and retrieval system, without permission from the Publisher in writing.

Typeset in 11/13pt Palatino
by Mac Style, Beverley, E. Yorkshire

Printed and bound in Great Britain
by MPG Books Group.

Pen and Sword Books Ltd incorporates the imprints of Pen and Sword Aviation, Pen and Sword Maritime, Pen and Sword Military, Wharncliffe Local History, Pen and Sword Select, Pen and Sword Military Classics and Leo Cooper.

For a complete list of Pen & Sword titles please contact
PEN & SWORD BOOKS LIMITED
47 Church Street, Barnsley, South Yorkshire, S70 2AS, England
E-mail: enquiries@pen-and-sword.co.uk
Website: www.pen-and-sword.co.uk

Contents

Foreword 6
Acknowledgements 8
Glossary 10
   1 Introduction 15
   2 Politics and Policy 21
   3 Equipment and Technology 53
   4 Personnel and Training 92
   5 Command and Control 134
   6 Concept and Doctrine 172
   7 Conclusion 200
Bibliography 205
Notes 216
Index 245

Foreword

Lieutenant General Sir Hew Pike, KCB, DSO, MBE

Military thinking tends towards conservatism, with professional soldiers seen by many as 'fighting the last war' in their approach to training and doctrine. But this is less true today than perhaps it has been in the past, and far more damaging to future military effectiveness are the attitudes of politicians and treasury officials towards the resourcing of defence capacity in a nation 'at peace' – attitudes which are strongly reinforced by the seemingly unassailable bulwarks of bureaucracy.

It therefore generally takes operational imperatives or a war to unblock things, and even then important developments remain prey to conflicts of interest, differing priorities and, not least, human prejudice.

Typical of such realities is the story of the development of Airborne Forces during the Second World War, and these difficulties are vividly analysed in John Greenacre's excellent study of what he aptly describes as 'Churchill's Spearhead'. Indeed, as he takes us through the thickets of politics and policy, of organisational and structural disagreements, of aircraft suitability and availability, of equipment and technical challenges, of manning, selection and training, of issues bearing upon command and control, and of conceptual and doctrinal debates, the wonder is that Airborne Forces ever got off the ground at all.

But launched they were, along with not only similar German and Russian capabilities (both, significantly, well ahead of the game), but also with those two great American Airborne Divisions, the 82nd and 101st – fighting with such distinction in Normandy with the British 6th and on Operation Market Garden with the British 1st Airborne Divisions. The author takes us through the setbacks and successes of early operations, and the manner in which these fledgling parachute and glider-borne forces found themselves, at length, part of something on a grander scale. Through it all, however, he never lets us forget the human factor that is central to warfare, whether it be in the bloodletting of the committee room or the struggles of the battlefield.

So there is much in this book that is highly relevant to our own situation seventy years on, where the same kind of debates about the future must be fought and won, so that military success can be delivered on operations. And let no man peddle that old cliché about Airborne Forces 'going from secret weapon to anachronism in one generation'. On the contrary, although this book finishes with the crossing of the Rhine in 1945, the role of British Airborne Forces in subsequent campaigns is vindication enough, if such were needed, of the inspired and utterly single minded determination of the early pioneers portrayed in these pages, and of the gallant soldiers who first proved the route down which their successors have so splendidly followed. Here, indeed, was born a band of brothers for all seasons, forged in the fires of a World War, developed through constant operations in the second half of the twentieth century, and now applying the skills and qualities of its soldiers to the complex, uncertain and dangerous environment of the twenty-first.

Utrinque Paratus
Hew Pike
November 2009

Acknowledgements

Acknowledgement is first of all due to Professor John Gooch who has guided, nurtured and encouraged me throughout my research and to Professor John Childs who has also provided valuable guidance and constructive criticism. I have also benefited from the patient advice and assistance provided by Maria Di Stefano.

I would have been unable to complete my thesis without the generosity and understanding of successive commanders who have allowed me the time to pursue my research, namely Brigadier Paul Cort, Brigadier Nicholas Eeles and Colonel David Turner.

I would like to thank the veterans of Britain's Second World War airborne community who have taken the time to reply to my enquiries and provide valuable and unique information and opinions, in particular Major General Tony Deane-Drummond, Lewis Golden, Peter Wilkinson, Arthur Shackleton, Jim Wallwork, David Brook, Robert Brown, Harry Howard, Herbert Buckle, Kenneth Frere and Bill Angell. Their help has been illuminating and important; any misinterpretation of their words or errors of fact are my fault alone. The patience and knowledge of the staff at the National Archives and the Liddell Hart Centre for Military Archives has been invaluable, as has the assistance of Tina Pittock at the Airborne Forces Museum in Aldershot, Jon Baker at the Airborne Assault Museum at Duxford and Derek Armitage and Susan Lindsay at the Museum of Army Flying at Middle Wallop. I am also grateful to the Imperial War Museum, Airborne Assault Museum and Museum of Army Flying for permission to reproduce photographs within this book.

I am indebted to the colleagues who have given their time to read and review my draft work and provide me with comments, suggestions and advice – Major Mike Peters, Major Steve Elsey and Major Andrew Roe amongst others. In addition I would like to acknowledge the support of my colleagues in the International Guild of Battlefield Guides. I also extend my sincere thanks to Sir Hew Pike KCB, DSO, MBE for taking the time to write the foreword for this book.

ACKNOWLEDGEMENTS

Finally and most sincerely I would like to thank my wife Jo-anne who has read every word of this book more than once and who has endured my late nights and days away and who has stayed with me despite my best efforts to the contrary. Also thank you to Alice who has put up with Daddy reading when he really should have been paying more attention.

John Greenacre
Pakefield, Suffolk

Glossary

Abbreviations

AAC	Army Air Corps.
AC Comd	Army Cooperation Command.
ACAS	Assistant Chief of the Air Staff.
AFEE	Airborne Forces Experimental Establishment.
AFHQ	Allied Force Headquarters.
AMSO	Air Minister for Supply Organisation.
ATI	Army Training Instruction.
ATM	Army Training Manual.
AFE	Airborne Forces Establishment.
AOC-in-C	Air Officer Commander in Chief.
AORG	Army Operational Research Group.
APTC	Army Physical Training Corps.
BAC	British Aircraft Commission.
Bde	Brigade.
BFTS	Basic Flying Training School.
BGS	Brigadier General Staff.
CAS	Chief of the Air Staff.
C-in-C	Commander in Chief.
CIGS	Chief of the Imperial General Staff.
CLE	Central Landing Establishment.
CLS	Central Landing School.
COC	Combined Operations Command.
COCHQ	Combined Operations Command Headquarters.
COS	Chiefs of Staff.
CRD	Chief of Research and Development.
DCO	Director(ate) Combined Operations.
DGE	Director General Equipment.
DMC	Director(ate) Military Cooperation.
DMO	Director(ate) Military Operations.
DMO&I	Director(ate) Military Operations and Intelligence.

Glossary

DMO&P	Director(ate) Military Operations and Plans.
DMT	Director(ate) Military Training.
DOR	Director(ate) Operational Requirements.
D Plans	Director(ate) of Plans.
DSD	Director(ate) Staff Duties.
DSR	Director(ate) Scientific Research.
DTD	Director(ate) Technical Development.
DZ	Drop Zone.
EFTS	Elementary Flying Training School.
FOPS	Future Operational Planning Staff.
GOC	General Officer Commanding.
GOTU	Glider Operational Training Unit.
GPR	Glider Pilot Regiment.
GTS	Glider Training School.
HGCU	Heavy Glider Conversion Unit.
ISTDC	Inter-Service Training and Development Centre.
JPC	Joint Planning Committee.
JPS	Joint Planning Staff.
L of C	Lines of Communication.
LZ	Landing Zone.
MAP	Minister (Ministry) of Aircraft Production.
MO4	Military Operations Department No. 4.
MOD	Ministry Of Defence.
MTP	Military Training Pamphlet.
PIAT	Projector Infantry Anti-Tank.
PTS	Parachute Training School.
RA	Royal Artillery.
RAF	Royal Air Force.
RASC	Royal Army Service Corps.
RE	Royal Engineers.
RM	Royal Marines.
RV	Rendezvous Point.
SAS	Special Air Service.
SASO	Senior Air Staff Officer.
SD4	Staff Duties Department No. 4.
SEAC	South East Asia Command.
SHAEF	Supreme Headquarters Allied European Forces.
SOE	Special Operations Executive.
SOP	Standard Operating Procedure.
SS	Special Service.
STO	Senior Technical Officer.
TDS	Technical Development Section.
TEWT	Tactical Exercise Without Troops.
USAAF	United States Army Air Force.

VCAS   Vice Chief of the Air Staff.
VCIGS   Vice Chief of the Imperial General Staff.
WO   War Office.

Additional Abbreviations Used in Footnotes
ABFM   Airborne Forces Museum.
AIR   Air Ministry Papers (National Archives).
AVIA   Ministry of Aircraft Production Papers (National Archives).
CAB   Cabinet Office Papers (National Archives).
DEFE   Minister/Ministry Of Defence Papers (National Archives).
IWM   Imperial War Museum.
LHCMA   Liddell Hart Centre for Military Archives.
MAF   Museum of Army Flying.
NA   National Archives.
PREM   Prime Minister's Papers (National Archives).
T   Treasury Papers (National Archives).
WO   War Office Papers (National Archives).

Military Operations referred to in the text
AVALANCHE   The amphibious landings and associated operations by United States Fifth Army at Salerno – 9 September 1943.
AXEHEAD   Planned and cancelled airborne operation to assist 21 Army Group to bridge the Seine – August 1944.
BEGGAR   Operation to ferry gliders from England to Morocco – 3 June to 7 July 1943.
BENEFICIARY   Planned and cancelled airborne operation to capture St Malo – 22 June to 3 July 1944.
BITING   Raid by a company of 2 Parachute Battalion to recover components of a German radar at Bruneval, France – 27 to 28 February 1942.
BLAZING   Planned and cancelled airborne operation to capture Alderney – May 1942.
BOXER   Planned and cancelled airborne operation to capture Boulogne and destroy V1 launch sites – August 1944.
COLOSSUS   Parachute raid by No. 2 Commando to destroy the Tregino aqueduct, Italy – 10 February 1941.
COMET   Planned and cancelled airborne operation to capture bridges over the Rhine from Arnhem to Wesel – September 1944.

Glossary

CORKSCREW	Operation to capture the island of Pantellaria in the Mediterranean – October 1942 to June 1943.
DEADSTICK	*Coup de Main* airlanding operation to seize bridges over the River Orne and Caen Canal on D-Day – 6 June 1944.
DRACULA	The British and Indian attack on Rangoon – 1 May 1945.
DRAGOON	(Originally Operation ANVIL). Allied operation to invade southern France – 15 August 1944.
FRESHMAN	Glider raid to destroy the Norsk Hydro plant in Vermork, Norway – 19 November 1942.
HANDS UP	Planned and cancelled airborne operation to capture the area of Quiberon Bay in Brittany – 15 July to 15 August 1944.
HASTY	A company airborne operation by 6 Parachute Battalion to harass German withdrawal east of Rome – 1 to 7 June 1944.
HUSKY	The Allied invasion of Sicily – 9 July to 17 August 1943.
INFATUATE	Operation to capture the Walcheren Islands – September 1944.
JUBILEE	Allied raid on Dieppe – 19 August 1942.
LANCING	Planned and cancelled airborne raids on Boulogne and Le Touquet – 23 to 25 May 1942.
LINNET	Planned and cancelled airborne operation to seize a firm base area near Tournai – September 1944.
MALLARD	Airlanding operation to deploy 6 Airlanding Brigade on the evening of D-Day – 6 June 1944.
MARKET GARDEN	Allied operation to drop two United States and one British airborne division in Holland and to link up using ground forces – 17 to 26 September 1944.
NEPTUNE	The Allied assault and amphibious operations as part of Operation OVERLORD on D-Day – 6 June 1944.
OVERLORD	Allied operations in France from D-Day – 6 June 1944 to the liberation of Paris – 25 August 1944.
PLUNDER	Deliberate Allied Rhine crossing operations – 23 to 24 March 1945.
ROUNDUP	A 1942 plan for the Allied invasion of northern France in the spring of 1943.
RUTTER	Original name used during planning for the Dieppe Raid, subsequently changed to JUBILEE.
SLAPSTICK	Landing by sea of British 1st Airborne Division at Taranto, Italy – 9 September 1943.

SLEDGEHAMMER	Allied contingency plan for a limited cross-Channel invasion in response to a German or Soviet collapse in 1942.
SWORDHILT	Planned and cancelled airborne operation to capture the Morlaix Viaduct – 20 July to 4 August 1944.
SYMBOL	Allied conference held in Casablanca – 14 to 24 January 1943.
THURSDAY	Allied 'Chindit' operation in Burma, March 1944.
TONGA	Operations by 3 Parachute Brigade and 5 Parachute Brigade on D-Day – 6 June 1944.
TORCH	(Originally Operation GYMNAST). The allied invasion of French North Africa – 8 to 10 November 1942.
TRANSFIGURE	Planned and cancelled airborne operation to secure the Paris–Orleans gap to cut off the enemy's retreat – 7 to 17 August 1944.
VARSITY	British and United States airborne element of Operation PLUNDER, the deliberate Allied Rhine crossing – 24 March 1945.
VERITABLE	21 Army Group operations to clear the land between the rivers Roer and Rhine – 8 February to 11 March 1945.
WILD OATS	Planned and cancelled airborne operation to capture Carpiquet airfield and Evrecy to encircle Caen – June 1944.

CHAPTER ONE

Introduction

British airborne warfare reached its zenith on 24 March 1945. Operation VARSITY dropped 6th British Airborne Division alongside 17th US Airborne Division as an integral part of Second British Army's wider Rhine crossing, Operation PLUNDER. The British airborne division, nearly 12,000 men with their weapons, jeeps, artillery, anti-tank guns and even tanks were carried in a single lift in 683 aircraft and 444 gliders. The men and equipment were carried over the heads of the assault troops crossing the river and were put down on their drop zones (DZ) and landing zones (LZ) east of the Rhine, around the Diersfordt forest and the village of Hamminkeln. Despite intense anti-aircraft fire the entire Division was on the ground within one hour of the first paratrooper exiting his aircraft. Four hours later all the Division's objectives were secure and the link up with the lead elements of Second British Army had been achieved.[1] For less than 300 men killed, 6th Airborne Division had neutralised a portion of the German indirect fire threat to the troops exposed in their assault boats and on the banks of the Rhine. It had also prevented enemy reserves from being brought up to interfere with the crossing, and by securing bridges over the River Issel, had created a bridgehead within which General Sir Miles Dempsey's men could establish themselves then push further east.

This level of military effectiveness is all the more remarkable when it is considered that the British military establishment had no concept of airborne warfare just a decade before. In 1935 the potential of the airborne assault was graphically demonstrated during Soviet Army manoeuvres outside Kiev. European ambassadors and military representatives, including Major General Archibald Wavell, witnessed the simultaneous drop of 1,500 armed paratroops. The airborne force captured bridges over the River Dnepr, 40 kilometres beyond the simulated front line, in order to prevent an imaginary enemy from reinforcing their forward echelon. It was an imposing illustration of the dawn of a new age of warfare. Major General Sir Hastings Ismay, responsible within the War Office for military intelligence relating to Russia at the time, was under no illusions that this was not a publicity stunt and that 'the Soviet Forces were making great strides in equipment and training and that... parachute units were receiving special attention.'[2] Wavell was not as impressed nor was he particularly enthusiastic about the opportunities that the Kiev exercise had demonstrated. 'This parachute descent, though its tactical value may be doubtful, was a most spectacular performance,'[3] he recorded at the time. On his return to England Wavell was appointed General Officer Commanding (GOC) 2nd Division in Aldershot. He gave a lecture to the officers under his command concerning the airborne exercise he had witnessed, closing with the words, 'I advise you when you go home to forget all about it.'[4]

A month after the exercise, on 26 October 1935, a captioned photograph of the Soviet parachute drop appeared in the 'Daily Telegraph'.[5] On the same day Major General John Dill, as Director of Military Operations and Intelligence (DMO&I) opened the War Office's official interest in airborne forces. On being shown the photograph, in a hand written note to the Director of Military Training (DMT) he observed, 'This is not the first we have heard of it and I feel that the time has come when we should do some experimental training on these lines.'[6] The Director of Staff Duties (DSD) agreed that the concept warranted scrutiny and so all three directorates of the War Office were united on the subject. Unanimity did not however lead to the project being vigorously pursued. Information was gathered concerning the state of airborne forces in France and Germany as well as in the Soviet Union. Dill discussed the theory of airborne warfare with Soviet Marshal Tukhachevskii during the latter's visit to London in 1936. All that the enquiries revealed was that Britain was woefully behind the other major European powers, both in conceptual and physical terms.[7] There was early recognition within the War Office that the Air Ministry must be part of the dialogue if the policy was to be moved forward and yet it took the Air Ministry nearly a year to respond to the War Office's request for the department's views on the subject. Once the Air Ministry joined the debate it became clear that any progression of concept and policy within the War Office would quickly be tempered, stopped or even reversed by the limited lift capacity of the Royal Air Force.[8]

It has been suggested that Dill was prominent in influencing the formation of both British and American airborne forces.[9] If this was the case then he certainly did not manage to convince his superiors of the worthiness of the concept during these pre-war years. Successive Chiefs of the Imperial General Staff (CIGS) (Field Marshal Sir Archibald Montgomery-Massingberd, Field Marshal Sir Cyril Deverell and Field Marshal Lord Gort) doubted the value of pursuing the project and so by mid-1938 airborne forces within the British military establishment were a purely theoretical group of units, with notional doctrine and tactics, to be used during tactical exercises without troops (TEWT) as a basis for further study. Any practical realisation was to be limited to experimentation only and even the benefit of this narrow pursuit was questioned.[10] With diplomatic and fiscal pressure being applied to the British armed forces during the 1930s and with no imperative function for such a capability it is unsurprising that airborne forces had been afforded such a low priority. Any idea of pursuing airborne development essentially ended in early 1938 with no tangible progress having been made during the previous three years. The project remained dormant until Germany used its airborne forces to such clear effect in Norway, Belgium and The Netherlands in April and May 1940.

So the British military establishment found itself at a standing start when it was ordered to begin airborne development in June 1940. It is therefore perhaps remarkable that the effect achieved during Operation VARSITY was at all possible less than five years later. Notwithstanding this accomplishment, historians in the intervening years have doubted the military effectiveness of Britain's airborne forces throughout the Second World War. This criticism has been centred on the premise that overall British airborne performance and achievements did not justify the investment of resources required to develop the airborne force. It has been argued that the airborne arm drained valuable manpower away from the hard-pressed infantry for no commensurate effect.[11] Marshal of the Royal Air Force Sir John Slessor suggested that the cost in terms of multi-engined aircraft, crews and RAF ground personnel committed to airborne warfare was not justified by results and that greater dividends might have been produced by similar investment in other forms of warfare.[12] Field Marshal Lord Carver believed that the effect of an airborne force on the conduct of operations was rarely as decisive as hoped.[13] These comments have clearly not been provoked by an examination of the effect achieved during Operation VARSITY but by a wider survey of the development and military effectiveness of British airborne forces during the Second World War. For every success, such as the spectacular operation to seize the crossings over the River Orne and Caen Canal in Normandy on D-Day, there were prominent failures and excessively costly enterprises.

The most obvious example of poor military effectiveness was witnessed during Operation MARKET GARDEN in The Netherlands in September 1944. 1st British Airborne Division was dropped near the town of Arnhem in order to seize and hold a bridge over the River Neder Rijn that would allow the rapid advance of Dempsey's Second British Army north to the Zuider See. Only a fraction of the fighting power of the Division ever reached the vital bridge and the airborne troops were forced to relinquish it before the ground force could arrive. The Division clung grimly to its bridgehead on the north bank of the river until it was forced to withdraw eight days after its initial deployment. When the Division managed to regroup only approximately 2,100 men and practically no equipment remained. Over 7,000 men were lost during the abortive operation. Fourteen months earlier the same Division had been part of the invasion of Sicily, Operation HUSKY. Although the airborne formation did manage to secure its objectives the accuracy of its glider landings in particular was woeful, and hence casualties in certain areas were excessively high. In some cases as few as 34 per cent of the gliders employed landed on their designated LZs, with over 50 per cent ditching in the Mediterranean.[14] This in turn led to the high casualty rate in the airlanding brigade of nearly 500 men killed, drowned, wounded or missing. In total the Division suffered over 700 casualties during the operation.[15]

The high casualties and relatively poor performance of these two operations were justified at the time by their influence on the wider operations of which they were part. In Sicily General Bernard Montgomery, the commander of Eighth Army during HUSKY, considered that the operations executed by 1st Airborne Division had accelerated his advance by seven days.[16] Post MARKET GARDEN, Montgomery again pronounced the operation 90 per cent successful. Certainly 1st Airborne Division's dogged battle at Arnhem prevented the Germans from bringing reserves to bear against 82nd US Airborne Division that was fighting to seize and hold the massive bridge over the River Waal at Nijmegen.[17] This bridge was secured and subsequently held by the allies, providing an important salient into the German-held Netherlands. This salient later became the start point for the first of those operations that cleared the west bank of the Rhine, Operation VERITABLE. So it would appear that from early 1943 until March 1945 the investment in development of Britain's airborne forces, despite fluctuations in military effectiveness, was considered worthwhile.

What is perhaps more difficult to justify is the return on the investment in airborne forces during the first half of the war. Britain's airborne forces were conceived by prime ministerial edict on 20 June 1940. From that moment until the beginning of November 1942 only two small-scale parachute raids were mounted, at Tragino in Italy (Operation COLOSSUS), and in Bruneval in France (Operation BITING). COLOSSUS was a partial success in that although the objective was achieved the entire party of paratroopers was captured. BITING was highly successful in seizing components of a German radar and bringing them back to Britain. Even so, these two operations, despite considerable investment in terms of manpower, equipment, infrastructure and staff effort represented a total return of just over 150 men committed to operations over a period of 29 months. Far from deriving maximum combat power from available resources, Britain's airborne forces were plainly not a militarily effective capability during this period.[18]

What is clear therefore is that the progress of British airborne warfare from the inception to the peak was not linear. Development proceeded at different rates during different phases of the process. Also, although the military effectiveness of Britain's airborne forces was undoubtedly high during VARSITY, it fluctuated widely along the path to that point. What then were the causes of this inconsistency in the developmental process and the wide variance in military effectiveness across the wartime period?

In 1948 Marshal of the Royal Air Force Sir John Slessor believed that 'the airborne forces of the Army won wide and well-earned fame in the late war. But I am not sure they have yet been the subject of all the clear and unemotional thinking they deserve.'[19] While no longer wholly true, sixty years later this statement still largely applies. There is now an extensive collection of literature linked to British airborne forces during the Second World War but, on the whole, it does not apply 'clear and unemotional thinking' to those questions concerning developmental progress and military effectiveness.

Much of the literature concentrates on the performance and experience of units, formations and individuals during airborne operations. However, there were extended periods of time when no British units were committed to airborne operations during the war, for example the eleven months between HUSKY and OVERLORD. Prior to November 1942 Britain's airborne capability was committed to just two operations totalling four days and involving approximately 150 men and yet the periods of development and training between operations have received very little attention. This has led to a bibliography where the demonstrable military effectiveness of Britain's airborne forces is relatively well documented but the immediate factors and underlying trends in development that influenced performance on any given operation have seldom been fully described or examined. As an example, 6th Airborne Division's highly successful operations on D-Day have been extensively documented. Likewise 1st Airborne Division's failure to achieve its objectives at Arnhem during MARKET GARDEN has an extensive library all of its own. What is not so readily apparent is any literature that attempts to draw a link between the two events. What could account for the disparity in military effectiveness between D-Day and MARKET GARDEN only three months later? Perhaps it can be attributed to differences in training methods adopted by the two formations or to a change in tactical doctrine between the two operations. Differences in the manner in which the two divisions were equipped could have had an effect, or individuals' command and control style might have been a factor.

Military innovation and the development of military capability do not generally progress smoothly along a defined chronological path. Therefore there is little value in attempting to examine the subject purely chronologically. Military capability is developed simultaneously across a number of distinct and separate lines, each of which may progress at a different rate to the others. In order to produce a full account, airborne development has to be examined as a number of separate lines of development.

Britain's airborne forces were conceived by political *diktat* as opposed to any clearly expressed military requirement. Therefore an important factor to be examined is the political influence on the creation of military capability during the Second World War, including inter-ministerial relationships and the influence of individuals in political power. The most obvious influence on military effectiveness comes from the physical potential and limitations of the capability itself, i.e. its constituent manpower and equipment. That influence is often expressed simply in terms of the raw numbers of men, guns or aircraft committed to a given operation. However, soldiers require training, and the rate and quality of that training is critical, as is the initial selection process in the case of specialist troops such as paratroopers and glider pilots. Equipment only reaches the battlefield after a lengthy acquisition and procurement process, which in turn is inextricably linked to the political environment.

This physical component gives a military capability the means to fight. However, the manner in which it fights is governed by moral and conceptual factors. Command and leadership is the point at which an individual has the most potential to influence the military effectiveness of an entire formation or establishment. The selection of commanders is paramount in a small, specialist organisation such as Britain's airborne forces. At the higher level airborne formations were directed and employed by corps and army commanders whose experience of airborne warfare, until the latter stages of the war, was extremely limited if not nonexistent. Combat officers of all ranks are, in theory, influenced in the decisions they make by the doctrine that has been imparted to them. While the development of tactical doctrine is often an iterative process fed by the lessons acquired during operations and training, a concept of the purpose and function of a military capability should be a more enduring statement, designed to guide development as well conduct during operations. Such a concept broadly existed prior to D-Day, appears to have been lost or ignored during MARKET GARDEN and was then reasserted for VARSITY.

Each of these lines of development – politics and policy, equipment and technology, personnel and training, command and control and concepts and doctrine – has an influence on all the others. They have to be considered together as well as individually in order to gain a more complete view of the process of airborne development. Once that is achieved the developmental path can be mapped onto a timeline of British airborne operational experience. Using this analysis it is possible to identify the link between the process of development and the military effectiveness of Britain's airborne forces during the war. We can then determine how, despite resource constraints and limited experience, British airborne warfare developed over a period of less than five years to the point where it could ultimately achieve its undeniable apogee of military effectiveness on the banks of the Rhine in March 1945.

CHAPTER TWO

Politics and Policy

Churchill and his Requirement

Churchill's arrival as prime minister must have been a shock to the system for many of those comfortably ensconced in Whitehall and its environs. As Ismay commented:

> The change in leadership may have given rise to a few misgivings in Whitehall. There is a type of senior official, both civil and military, who get more and more set in their ways as they ascend the ladder of promotion. These able, upright, worthy men do not like the even tenor of their lives disturbed, and resent dynamic ministerial control. This is precisely what they were likely to suffer at Churchill's hands.[1]

Even the most capable, willing and flexible officers and civil servants began to feel the strain soon after the change of administration. The Prime Minister's working hours were unconventional, his output prodigious in both quantity and scope. He was interested in the minutest details of everything the Staff did and he poured out floods of memoranda upon a plethora of problems.[2] To cope with the sheer volume, the validity of Churchill's queries and directions had to be interrogated in order that some priority could be brought to the work required. This is perfectly normal practice in any organisation where staff are under to pressure to deliver results within the constraints of available time. It is particularly apparent when the head of that organisation has a reputation for burdening his workforce with tasks that are unachievable, outlandish or irrelevant, intermingled with those which are vital. Slessor, on the Air Staff for much of the war as Director of Plans, wrote of the Prime Minister's hand in the inception of British airborne forces that although Churchill's indomitable offensive spirit was a tonic to those surrounding him it was liable to manifest itself 'often quite regardless of practical realities.'[3] Dill concurred following his appointment as CIGS in April 1940, 'He [Churchill] is full of ideas, many brilliant, but most of them impracticable.'[4] Consequently those who were the immediate recipients of the Prime Minister's directives might be forgiven if, from time to time, some of those that appeared less relevant or only remotely obtainable were sifted, either consciously or unconsciously, into the category of work that might safely be left until tomorrow.

The new Prime Minister, along with most of the high command at the time, had been deeply impressed by Germany's airborne campaign in Norway, the parachute assault on The Hague and the glider enabled capture of Eben Emael during 1940. On 22 June Churchill ordered that Britain should begin the development of its own airborne capability.

> We ought to have a corps of at least 5,000 parachute troops, including a proportion of Australians, New Zealanders and Canadians, together with some trustworthy people from Norway and France. I see more difficulty in selecting and employing Danes, Dutch and Belgians. I hear something is being done already to form such a corps, but only I believe on a very small scale. Advantage must be taken of the summer to train these forces, who can, none the less, play their part meanwhile as shock troops in home defences. Pray let me have a note from the War Office on the subject.[5]

It was not difficult for anyone to assess the offensive potential of an airborne arm; it was, however, difficult in June 1940 to foresee a time when Britain might be in a position to employ it. Whilst Churchill might have had the vision to anticipate fighting his way back into continental Europe, he did not assist those around him by precisely articulating his appreciation. His minute of 22 June 1940 lacked, in several areas, the definition required to enable his intentions to be converted into reality. Those few lines sacrificed clarity for succinctness. The vagueness of his request no doubt decreased the probability of any constructive reaction. The Air Ministry summed up the situation: 'This requirement in itself is insufficient to enable all concerned to go ahead satisfactorily with the production of such a [airborne] force.'[6]

The Prime Minister's statement gave no indication of how 5,000 parachutists might be incorporated within a formation or what the ultimate purpose of that formation might be. Even the figure itself is confusing: 5,000 does not equate to any contemporary fighting unit and falls between the fighting establishment of a conventional brigade and a division. There is no mention of gliders, despite Germany's success employing this capability, an omission that was to have enduring consequences as the airborne concept was devised and amended. The explicit inclusion of foreign nationals in the minute also had the possible effect of introducing confusion. Why would the Prime Minister be so specific about training these soldiers, particularly those from occupied Europe, unless he had a distinct task in mind for them?

Only a month before, on 19 May 1940, the Chiefs Of Staff (COS) Committee had considered a report entitled 'British Strategy in a Certain Eventuality'. The eventuality being the fall of France and the strategy one advocating economic warfare and a bombing offensive together with the use of resistance movements to ignite widespread insurrection throughout occupied Europe. In 1940–41 the 'detonator concept' was part of Britain's long-term strategic hopes for independent victory over Germany.[7] There was a school of thought in 1940 that this type of warfare was ungentlemanly and therefore not suitable for regular British soldiers to be involved in. Clement Attlee, the Lord Privy Seal, recommended that it was the nationals of the oppressed countries who should be the direct participants in any form of insurrection.[8]

It is probable that Churchill's minute of 22 June became linked with the strategic paper of 19 May in some quarters; after all what better way of returning trained guerrillas to their homeland than by parachute. There is evidence for this as on 4 September 1940 the Directorate of Military Operations (DMO) suggested that the facilities at Ringway (the establishment responsible for parachute training) would best be employed in training Poles, Czechs and other selected foreigners, who could lead and foster rebellions in their own countries. It was believed that this would follow the example set by Russia by concentrating on the parachute training of specialist individuals rather than formed military units.[9] The Central Landing School (CLS) at Ringway was the only establishment formally training parachutists in Britain in June 1940 but, as alluded to in Churchill's minute, its throughput was only on a 'very small scale'. If the meaning of the minute was interpreted so as to assume that the insertion of insurrectionists was the Prime Minister's aim then this could be met by simply increasing the training rate at the CLS, with very little effort or original thought being required. There is little evidence outside of the DMO note that this train of thought existed but it can be understood how it might be taken as an excuse for inaction.

Had Churchill closely monitored his instructions during their implementation then any confusion might have been easily rectified with a few words of clarification direct from the author. Instead the history of the Prime Minister's attitude and focus towards this project, that some considered was very largely his own personal 'bee', is erratic and inconsistent.[10] The first member of the Staff to engage directly in response to the initial minute was the Director of Combined Operations (DCO), Admiral of the Fleet Sir Roger Keyes. He had quickly grasped that the provision of suitable aircraft in sufficient numbers would be a crucial factor and on 27 July 1940 he urged Churchill to use his influence to secure the use of Douglas transport aircraft from both the American and Dutch Governments.[11] The go-ahead was given to Ismay to staff the matter. Within the same letter Keyes mentioned that of 3,500 volunteers for parachuting 500 had been selected for training. Churchill ringed the latter figure and noted in the margin, 'I said 5,000.' This fact was reiterated at the 250th meeting of the COS Committee on 6 August, at which Keyes was present, and Churchill repeated the same comment in the margin of the minutes, this time underlining it.[12] Ismay pointed out that while 5,000 was still the ultimate target, 500 represented an intermediate aim and was more realistic in the short term.[13]

Later the same month the Air Staff submitted a report dismissing the potential of paratroops in favour of pressing ahead with glider development. On 1 September the Prime Minister responded to the Air Ministry by stating that obviously glider development should be pursued if it represented a better capability than parachutists. However he did question whether it was being taken up seriously, 'Are we not in danger of being fobbed off with one doubtful and experimental policy and losing the other which has already been proved?'[14] He requested a full report in typical fashion.

This he received from the Air Staff, via Ismay, a few days later. Ismay now decided to submit a minute bringing together those decisions already made concerning the training of paratroops and the Air Ministry's recent thoughts on glider development.[15] A set of figures was lucidly presented stating that any operation would require an airborne force of not more than 1,000 men of which only 100 would need to be parachutists, the rest being glider-borne. Taking into account multiple operations and the need for a reserve the total airborne requirement was stated as 3,200 men of whom 500 needed to be parachutists. 'Press on', Churchill wrote at the end of the note. With this comment he tacitly, and probably unintentionally, reduced his original target for an airborne force, reduced the training requirement for parachutists to only 500 and put airlanding troops ahead of paratroops in terms of priority, in spite of the fact that the gliders needed to deliver them were still largely confined to the drawing board.

The Prime Minister's enthusiasm and involvement now lapsed until the following spring. Having done little to supervise or even monitor the progress of development during the intervening months, Churchill visited the school at Ringway, by now renamed the Central Landing Establishment (CLE), on 26 April 1941 in order to watch an airborne demonstration. What he witnessed was a graphic display of the lack of progress that had been made over the past ten months. A formation parachute drop by six Whitleys (therefore a maximum of sixty men) was complemented by a formation landing by just five civilian, single-seat 'sailplane' gliders. A fly-past by a single Hotspur was the only demonstration of military gliders that could be mustered.[16] It was a sorry effort due in large part to the lack of aircraft being made available to the CLE at the time.

Despite the poor show Churchill was not unimpressed by what he had been shown but was depressed by the paucity of paratroopers and the sluggish pace of glider development. He left Ringway with the impression, given to him by the CLE staff, that the Air Ministry's apathy was to blame and, in the view of others, he began to believe that the Air Staff were deliberately thwarting his airborne project.[17] As will be seen, this belief was not without justification. On return to London Churchill reviewed the programme and on 27 May 1941 he sent a personal minute to the COS Committee via Ismay.[18] In it he blamed himself and the Air Ministry equally for the situation: the Air Ministry for offering wrongly based resistance to his initial requirement and himself for becoming overborne by that resistance. He referred back to his minute of 1 September the previous year and that his fears had been realised with neither a credible parachute training or glider development programme now in place. He impressed upon the COS Committee that the airborne capability was not, in his opinion, an expensive luxury and that it would be necessary for offensive operations in the Mediterranean and Middle East in 1942. He placed the problem squarely in their laps, 'A whole year has been lost, and I now invite the Chiefs Of Staff to make proposals for trying, so far as is possible, to repair this misfortune. The whole file is to be brought before the Chiefs Of Staff this evening.'

As might be expected, a flurry of activity followed. The COS Committee issued a Joint Memorandum to the Prime Minister on 31 May 1941. In it the War Office set their requirement at two brigades of parachutists, one to be based in the UK and one in the Middle East, and a glider organisation capable of carrying a brigade group, again to be duplicated in the Middle East. The Air Staff recognised that the provision of aircraft and gliders would require considerable effort in order to keep pace with the Army's aspirations. Nevertheless, a couple of months later, they were confident of being able to provide up to ten heavy and medium transport squadrons and modified bombers capable of lifting 2,500 parachutists in line with the War Office's development plan.[19] The ability to suddenly promise this support might suggest that Churchill's suspicion of deliberate hindrance from the Air Ministry had been justified. However, despite his minute earlier in the month the programme put forward by the COS Committee was not designed to produce the promised results until the summer of 1943 when, in their opinion, the strategic situation would be more conducive to the employment of an airborne force.[20] In spite of this the Prime Minister, as Minister of Defence, must have felt confident that the COS Committee was as good as its word and that the airborne programme was being developed as fast as possible.[21] Once again he loosened the reins and, aside from occasional correspondence with the India Office on the subject (which will be covered later in this chapter), he did little to monitor progress.

The War Office did now begin to make significant progress. The embryonic 11 Special Air Service (SAS) Battalion became 1 Parachute

Battalion and was then expanded to form 1 Parachute Brigade. An independent infantry brigade was converted to the airlanding role and an airborne divisional headquarters was approved and formed, all within six months. In September 1941 the Air Ministry renamed the CLE the Airborne Forces Establishment (AFE), increased its remit and expanded the organisation accordingly. Two permanent exercise squadrons (numbers 296 and 297) were formed under the new 38 Wing RAF in January 1942. Despite this effort, the RAF did not keep pace with the Army's expansion of airborne forces. 38 Wing had still not reached its establishment for aircraft by April 1942 and even if all the aircraft had been available the numbers in the two exercise squadrons would still have been inadequate.[22] There simply were not enough aircraft available to train the throughput of the AFE and to provide exercises and maintain the skills and readiness of those units already trained. This fact was recognised by the Secretary of State for Air, Sir Archibald Sinclair, who signalled the fact to the Ministry of Aircraft Production (MAP) at the highest level during March 1942.[23]

On 16 April 1942 Churchill paid a visit to 1st Airborne Division expecting to see a demonstration of its full capability. He was treated to a repeat performance of the display he had witnessed almost exactly a year before. Only twelve Whitleys were available, dropping a maximum of 120 parachutists.[24] In addition nine aging Hawker Hectors each towed and released a single Hotspur glider, of which one overshot the landing zone and another hit trees and crashed on its approach. Even those who took part could tell that Churchill was 'furious at the poverty of numbers' involved in the display and that 'fireworks' from the Prime Minister were likely to be the result.[25] Once again those 'fireworks' led to a flurry of staff activity. Major General Browning, GOC 1st Airborne Division, wrote a report to the Prime Minister on the same day as the lacklustre demonstration. In it he blamed a lack of drive and enthusiasm from the War Office and the Air Ministry for the shortages in personnel, aircraft, equipment and weapons, which had led to development being disastrously slow.[26] Both the General Staff and the Air Staff responded but Churchill was clearly unconvinced by their proposals and finally acted decisively to seize the situation.

On 1 May the Prime Minister instigated an Airborne Forces Committee and through MAP, Sir Robert Renwick, a prominent industrialist, was appointed Chairman.[27] In actual fact MAP had already created this committee. Following a letter from Sinclair, Colonel The Right Honourable J.J. Llewellin of MAP on 4 April ordered a Glider Committee to be formed with Renwick at the chair.[28] On 10 April Renwick's terms of reference were determined: they were 'to coordinate arrangements for the development, production, supply, transport and storage of all equipment for airborne forces, and to secure rapid decisions.'[29] The Airborne Forces Committee met for the first time on 24 April 1942 with representatives from the Air

Ministry, War Office, Ministry of Aircraft Production, the Airborne Division and 38 Wing RAF. It appears likely that, following the debacle on 16 April and the dearth of firm proposals to remedy the situation, the first meeting of the Airborne Forces Committee was brought to Churchill's attention as an example of positive action. He then gave the committee his official endorsement and Renwick direct access to himself before its second meeting on 1 May 1942. Although the committee was late in the day and not entirely of his own making, Churchill had at last used his position as Minister of Defence to enable 'firm decisions to be reached and translated into action far more quickly than had hitherto been the case.'[30]

It is unclear why the Prime Minister had not ordered a full committee to be formed in June 1940. He was certainly a proponent of this method of doing business, the Tank Parliament being a well-known example. The true advantage of Churchill's hand on the tiller became apparent when an issue involved more than one service. The Battle of the Atlantic Committee, set up to combat the U-Boat menace, coordinated the actions of the Royal Navy and the RAF. The Night Air Defence Committee desegregated the independent actions of all three services and had a synergistic effect as a result. The development of airborne forces would appear to have been an obvious candidate for the Prime Minister's personal attention as an invaluable 'progress-chaser', encompassing as it did the most acrimonious of inter-service rivalries at the very heart of the requirement.[31] Churchill's position as Minister of Defence was crucial as the final arbiter of policy with respect to the war effort. It was he who had the breadth of vision to sift and endorse the requirements being put forward by the Joint Planning Staff (JPS) and the authority to ensure that the COS Committee put the organisations, personnel and equipment in place to meet those requirements. Churchill's early and consistent intervention might well have 'removed bottlenecks and hastened growth.'[32] Instead it took the results of two years of apathy, lethargy and procrastination to produce any sort of decisive action from the Prime Minister.

Renwick's Airborne Forces Committee convened approximately fortnightly for nearly a year and made positive progress during that time. However, its remit still fell far short of that necessary to ensure smooth and expeditious progress in all the lines of development of the airborne force's capability. For reasons that will be seen later, India was an enthusiastic proponent of the airborne concept and had created its own Airborne Forces Committee a full year earlier on 22 April 1941.[33] It looked holistically at the problem of airborne development with separate sub-committees studying organisation, training and equipment. In contrast, Renwick's terms of reference restricted his committee's activity to solving the problems connected with the provision of aircraft and associated airborne equipment. In fact the committee deliberately avoided widening its scope. 'I do not think that you need fear that the inter-departmental committee, whose terms of reference include the phrase "Airborne Forces", will concern themselves with discussion of the wider questions of policy', Llewellin assured Sinclair on 20 April 1942.[34] So those wider questions were left to the COS Committee to tackle, despite the fact that the Air Staff and General Staff still had major differences of opinion concerning fundamental questions as to the future airborne forces' size, organisation, concept and doctrine.

Wrangling and jostling for position continued throughout the summer of 1942 without resolution ever appearing any closer. Finally, on 24 October Ismay was forced to raise the matter with Churchill. 'The Chiefs of Staff have had several discussions about airborne forces… CIGS and CAS [Chief of the Air Staff] find themselves, however, unable to reach any form of agreement. The Chiefs of Staff therefore have no alternative but to refer the matter to arbitration.'[35] The minute was accompanied by a fourteen-page paper by the COS Committee (COS (42) 434), signed off by CAS, Air Chief Marshal Sir Charles Portal, on 21 October.[36] Portal made it clear from the outset that there was no prospect of arriving at any agreed conclusion as to the future of Britain's airborne forces. He stated that it was plain that the two points of view had become so firmly entrenched as to be irreconcilable. The paper included an annex, 'The Value of Airborne Forces', from CIGS, General Sir Alan Brooke, and a report by the Air Officer Commanding-In-Chief (AOC-in-C) of Bomber Command, Air Chief Marshal Sir Arthur Harris, which will be studied in more detail later. The Minister of Defence was requested to make a decision that, if it went the Air Staff's way, would lead to a serious reduction in the striking power of the proposed airborne forces and the restriction of their role to minor operations only. No decision was forthcoming from Churchill. This caused consternation among the General Staff who believed that the Air Staff assumed that they had already won the argument and were acting accordingly. On 7 November Ismay wrote to Brigadier Richard Gale, Director of Air in the War Office, 'I do not wonder that you are worried about the delay in taking COS (42) 434, but it is a matter on which the Prime Minister alone [can] arbitrate, and he is terribly preoccupied at the moment. In point of fact I brought it to his notice last night, but without success.'[37] Churchill appeared unwilling or unable to adjudicate at a time when airborne forces were about to be committed to action, for the first time in appreciable numbers, as part of Operation TORCH in North Africa.

At the end of 1942 with the Prime Minister focussed on TORCH, Ismay finally drafted a response to the COS Committee on his behalf.[38] In it he made it clear that the Air Ministry position should prevail, noting that it was difficult to escape the conclusion that the size and organisation of the airborne force has been settled without full regard to its implications on the war effort as a whole. He went on to conclude that it was hard to see a point at which mass airborne forces would be required except during the execution of Operation ROUNDUP (later OVERLORD). ROUNDUP had already been put off until 1943, following America's acceptance of Operation TORCH over Operation SLEDGEHAMMER in 1942, and would be postponed again to 1944 by the SYMBOL conference in January 1943. Ismay suggested that any future airborne organisation should be limited in size to two parachute brigades and a small glider force of not more than two hundred gliders. Churchill assented to this view and requested Ismay to draft a minute accordingly for his consideration. In actual fact the minute issued to the COS Committee by the Prime Minister three days later bore no resemblance to Ismay's original draft. It was watered down, non-committal and gave no firm direction. Churchill felt instinctively that the airborne programme probably was overambitious but he put off translating his feelings into direction by ordering another review of the situation. He did not want the Chiefs of Staff to be unduly burdened with the question during a period of crucial operations. He believed it would be better if the Vice Chiefs gave it special attention, which should not have taken more than two sittings. He hoped that their report would give him something to work upon.[39] The Vice Chiefs had been appointed since 22 April 1940 in order to ease the burden on the Chiefs of Staff who had a dual role in advising the government on defence policy as a whole while at the same time directing the work of their own individual service. The Vice Chiefs of Staff were directed to hold regular meetings at which they would deal, in the name of the Chiefs of Staff Committee, with such matters as were delegated to them.[40] The stand-off over airborne forces was one such matter but such intractable problems are seldom solved through delegation.

At this point the Air Staff had their case for a reduction in airborne development bolstered by the support of another ministry. MAP was not always a leading proponent of Air Ministry policy but in this case, either through collusion or coincidence, it produced an authoritative and timely intervention. Portal's self-satisfaction can almost be detected when he wrote to Churchill on 14 November to inform him that 'I have just received an urgent message from M.A.P. to the effect that unless a stand-still order for gliders is given within the next two or three days they will be committed to the whole order of 2,000,' adding somewhat superfluously, 'You are familiar with the seriousness and magnitude of the implications of the glider programme.'[41] Churchill was now forced to make a decision and that decision, inevitably, had to favour the Air Staff's position. Reiterating that it was all a question of balance and emphasis, the Prime Minister issued a minute to the COS Committee on 17 November reflecting Ismay's original draft of 9 November.[42] Anxious that there was no prospect in the near future of the Air Ministry being able to provide the necessary aircraft for an airborne force of the size contemplated by the War Office, Churchill limited any future organisation to two brigades and a small glider-borne force, adding that an immediate halt to glider production should be ordered.

Notwithstanding Churchill's caveat that the whole position should be re-examined in six months' time, British airborne forces appeared to have been struck a serious, if not mortal, blow that would confine them to the periphery of operations for the remainder of the war. The matter appeared to be settled in the Air Ministry's favour. However British airborne forces were to receive a swift reprieve. Following SYMBOL, the allied conference that opened in Casablanca on 13 January 1943, Sicily was accepted as the next objective in the Mediterranean campaign. By early February 1943, although the precise requirements for the invasion (Operation HUSKY) were still being worked on, it was clear that considerably more airborne forces would be required than were at that time provided for.[43] The JPS's original plans for an invasion corps had included an airborne element.[44] The failure of the Dieppe Raid meant that any capability that might soften the impact of a frontal amphibious assault was seen as vital. The original plan for Dieppe, Operation RUTTER included an airborne element that was discarded in the final plan for JUBILEE. Despite Churchill's six-month moratorium it was clear that a re-examination could not wait and would have to take place at once. Following further protest, the Air Staff had this time to submit to the requirements of the planners. From this point on British airborne forces size, organisation, concept and doctrine would owe more to operational imperative and less to political expediency. By the beginning of June 1943 not only had 1st Airborne Division been reprieved in its original form but a second, 6th Airborne Division, had been added to the order of battle.[45]

Aside from the performance of the gliders and in particular their towing aircraft, those in authority who witnessed HUSKY agreed that airborne forces would play an essential role in any future invasion of the European mainland. A minute to the Prime Minister following the assault on Sicily carried an extract from a report by General Sir Harold Alexander:

> We must at once raise, organise, equip and develop an airborne force of parachutists and gliders – say, a corps of two divisions.... I know the answer will be that it is quite impossible to afford the pilots and the aircraft. Well! It is a question of priorities and personally I firmly believe that with our growing air supremacy, priority No.1 is for the Airborne Corps.[46]

The fact that Churchill was responsible for the vision behind the initial decision to form British airborne forces is not disputed. However, it needs also to be recognised that on more than one occasion he was also responsible for very nearly terminating the entire programme or at least curtailing it at a level where its impact would have been negligible. The full effect of his poor initial statement of requirement will be discussed further in this chapter as will his failure to initiate a standing committee with a chairman with the power to influence all the requisite lines of development. Clearly the Prime Minister could not have been expected to constantly follow every step of the programme as it evolved. However, at times he appears to have lost interest in his 'bee' altogether. Then, when matters were at their lowest, following the contemptible demonstrations in the spring of 1941 and 1942, he immersed himself in minutiae, a habit that frustrated those around him. Following the April 1941 display he wrote to Ismay, 'Parachutists who landed on Saturday had their knuckles terribly cut. Has the question of protecting their hands and kneecaps been considered?'[47] Ismay does not appear to have followed up the enquiry. Again after the shambles in April 1942, on the same day as the display he requested information from the War Office concerning airborne soldiers' extra pay.[48] Still more difficult to understand is how Churchill, who had a clear vision of airborne forces' future utility, allowed himself to be persuaded to make decisions on two occasions that would, to all intents and purposes, have brought a halt to development, first in September 1940 and then again in November 1942. Without a specific committee he had to rely on the judgement and advice of the Chiefs of Staff, men that Churchill considered 'the dead hand of inanition,' and 'ancient weapons' during the early part of the war.[49] The Chiefs, in turn, relied upon the judgement of their staff. The Prime Minister's decisions were therefore based on the advice of the General Staff and Air Staff. It was advice that was often based on the partisan agenda of a single service rather than the objective assessment of the requirements of the nation's war effort.

The General Staff and Air Staff and their Resistance
In addition to the faults with Churchill's initial requirement already outlined there was another, not yet alluded to, that was more pervasive and had more far reaching consequences than confusion over numbers or the employment of foreign nationals. Having not ordered the formation of an airborne forces committee, the Prime Minister also failed to designate who was to take the lead in development. He did not nominate a single ministry, office or department to coordinate the many lines of development that would be required to be brought together to produce an effective airborne capability. The minute of 22 June 1940 was addressed to the War Office via Ismay and therefore, perhaps, the selection of the Army as lead service was intended to be self-evident; after all, the decisive act of an airborne force would always take place on the ground. However, the controlling interest of the RAF in any airborne operation might have suggested that they could appropriately take the lead, as was the case with Germany's airborne forces. The solution was not obvious and the resulting dilemma became chronic.[50]

Churchill made the context from which a decision had to be extracted more difficult by designating a particular service, rather than one or more lines of operation, as the overall main effort for this early phase of the war. 'The Navy can lose us the war, but only the Air Force can win it.... The Air Force and its action on the largest scale must therefore... claim the first place over the Navy or the Army.'[51] This was only ever intended by the Prime Minister to be an indication of the priority of supply of munitions to the RAF in order, firstly, to achieve air supremacy over Britain, and secondly, to prosecute the bombing campaign over Germany. However, to some in the Air Staff it must have appeared to have given them *carte blanche* to swat away requests for support from the other two services. The result was friction between the Air Staff and the General Staff that resulted in the case for airborne forces being taken by the Chiefs of Staff to the Minister of Defence for arbitration following nearly two-and-a-half years of development, impeded, in part, by inter-service resistance.

That resistance took three forms. The first was straightforward, old-fashioned inter-service rivalry. A considerable amount of prejudice, distrust and lack of understanding had built up during the inter-war period.[52] The second was a constructive resistance, practiced by the Air Staff on two grounds: that the amount of effort expended on airborne forces would never produce a commensurate effect and that total commitment to the bomber campaign would produce quicker results. The third was a passive resistance, practised by the General Staff. This manifested itself in some parts of the War Office as a lack of belief in the potential of airborne forces and a corresponding deficiency in the staff effort dedicated to supporting the case for development. Each of these forms of resistance will now be studied in more detail.

A basic rivalry existed between the Air Staff and RAF and the General Staff and Army that was far from petty. It was 'a long and tortuous dispute between the two Services that dated back to the last years of the First World War', born out of a sincerely held mistrust of the intentions and actions of the other service.[53] The RAF had a precarious youth and during the austere inter-war period had to carve itself a niche from which it could defend itself against the perceived hostile intentions of the other services. Under the dominant influence of Sir Hugh Trenchard, the Air Staff developed their independent bombing doctrine and regarded any diversion from that concept as a threat to their survival.[54] Hence requests for support from the other services were looked on with suspicion. They still regarded the General Staff as 'wicked uncles who, although ostensibly reformed, might once again revert to predatory instincts.'[55] A wedge was driven between the two staffs, each of which had its own conception of future war.[56]

Having isolated itself from what it saw as attacks on its independence, the RAF developed it's bombing doctrine into dogma. Commanders began to believe their own publicity and 'luxuriated in the conviction "We are, ergo we are capable of a strategic bombing offensive.'"[57] This attitude was recognised within the Army, some of whom believed that the main trouble with the RAF was that since its inception it had been encouraged to imagine that it could and should win campaigns and wars through the application of air power alone.[58] There were others who considered the Air Staff vision of future warfare to be more accurate than those of the other two services. However the RAF's unyielding attitude towards its core doctrine of bomber supremacy and its corresponding ability to prosecute it remained open to question.[59] Due to lack of investment and the resultant paucity of effective modern aircraft the Air Staff's bombing doctrine was revealed for a set of emperor's clothes at the outbreak of war. To the General Staff it appeared that their years of prejudice had been justified. The relationship became increasingly strained; one result being that cooperation between the Army and the RAF was, to put it mildly, still in a rudimentary state in 1940.[60]

The situation improved as the war progressed but this was due to necessity and compulsion as much as to any greater degree of mutual appreciation. Although the schism may have become less publicly acceptable, mistrust still existed and not always without good cause. As late as 1943 there were still senior RAF officers who 'regarded every transport aircraft built at the expense of a bomber as a major tactical defeat.'[61] The Army's predatory instincts remained intact at the very highest level and continued to be revealed privately up until 1942 by CIGS.[62] Basic prejudices, although suppressed, continued to influence inter-service cooperation throughout the war. The manner in which this manifested itself in the case of the development of airborne forces differed markedly between the two services.

It would be unfair to suggest that the Air Staff had 'failed to recognise the principle that any theory or weapon of war is effective only if the means are available to exploit it appropriately.'[63] At the outbreak of war they were only too aware that appeasement and a dearth of defence spending had left their underpinning doctrine dangerously undermined. A twin track approach was taken to recover their position; an almost fanatical dedication to building up the bomber force while simultaneously denigrating any other concept that might encroach on that main effort. The vertical, functional command structure of the RAF exacerbated the results and all the RAF Commands resented any attempted incursion into their war effort, either from elsewhere within their own service or from outside.[64] The idea of using bombers to drop paratroopers and tow gliders was 'naturally repugnant to the Air Staff and to Bomber Command,' in particular.[65] As Churchill issued his minute in June 1940 the Air Staff were already fighting one political battle. Lord Beaverbrook had recently taken up the position of MAP and was advocating a drastic increase in fighter production at the expense of the bomber force. The Air Staff were attempting to counter this along the lines that the 'multiplication of fighters was a heresy which appealed only to those who were ignorant about air power.'[66] With the inception of airborne forces they were quick to state that no new commitments could be accepted without detracting from the previously approved expansion programme. If such commitments were accepted they would have to be at the expense of the future bomber force.[67] From this the Air Staff concluded that it was difficult to envisage a situation in which the number of bombers required could be spared from their normal task, to be risked in such a hazardous pursuit as an airborne operation.[68]

The argument was taken up at the highest level: 'Frankly I regard the bombing of German industry as an incomparably greater contribution to the war than the training and constant availability of an airborne division', CAS declared, 'and, as the two things at present seriously conflict, I would certainly accord priority to bombing.'[69] The zenith of the depletion of the bomber force as an argument against full development of airborne forces was reached in the paper presented to Churchill for arbitration in October 1942. In 2,000 words Air Chief Marshal Sir Arthur Harris presented the entire argument. His statement contained many lucid points against which it appeared difficult to argue; the training burden, the technical unsuitability of most bomber aircraft, the impact on the bombing campaign of the intensive training required immediately prior to an airborne operation, even the poor meteorological conditions prevalent in northern Europe. Harris's summarising remarks are worth reproducing.

> It would require the whole of the existing Bomber Command to be taken off operations for a period of four to six weeks… to transport one brigade for one operation; it would require about 2½ times the strength of the present Bomber Command to do the same for the airborne division as a whole…. The crux of the matter is this – is Bomber Command to continue its offensive action by bombing Germany, or is it to be turned into a training and transport Command for carrying a few thousand airborne troops to some undetermined destination for some vague purpose? There is no possibility of compromise…. Finally, I must record my conviction that had we sufficient air resources to transport an airborne force that could have any decisive influence on the outcome of the war, they would be sufficient to bring Germany to her knees by the simple process of carrying sufficient explosives for that purpose.[70]

It was this argument, in part, that finally persuaded Churchill to reduce the commitment to airborne forces and limit their size to two parachute brigades and a small glider force.

However, the Air Staff could not rely on the bomber supremacy argument alone, after all it had not defeated Beaverbrook in 1940.

Systematic criticism of the capability of airborne forces was also required, a task that the Air Staff took to with enthusiasm. Statements decrying the weak initial requirement and the equally weak War Office response were reasonable. It was undeniable that there had been no policy published governing the type of operations on which it was intended to employ the airborne force. Nor had the basic question of the proper composition of the airborne force been fully investigated.[71] The proposition that the airborne commitment had been accepted because those involved in the decision were not aware of the governing air factors and had therefore agreed to attempt impractical tasks on an unsound basis was also essentially fair. The supposition that there was little prospect of repeating in Europe the successful airborne operations executed by the Germans in 1940 was more tenuous and proved to be so.[72] Nevertheless the pressure was maintained and the perceived inadequacy of airborne forces continued to be briefed to and by the most senior representatives of the RAF. Slessor was ever sceptical: 'We do not know if we will ever have to use a [airborne] force of this nature, and certainly at the present time it would be wasting a lot of valuable effort to attempt to produce one.'[73] Even the War Office motives for wanting an airborne force were questioned by the very RAF command charged with supporting the Army. 'It is possible that the War Office insistence on putting in for these large requirements is to ensure that history may be able to record that they were not blind to modern developments!'[74]

The General Staff's response to this constant sniping was poor. The War Office was well aware that within the Air Ministry the atmosphere was distinctly unfavourable towards the provision of airborne forces.[75] The General Staff failed to respond in kind and did not counter the bomber supremacy line with any constructive argument of their own. Instead they criticised the Air Staff for constantly presenting limitations and restrictions. Describing the Air Ministry's lucid and persuasive staff work as 'very wet' can hardly be considered constructive.[76] It was also unhelpful when the highest ranked Army officers publicly displayed no faith in the airborne capability themselves: 'VCIGS stated that he could not visualise any substantial success for an isolated airborne invasion, anywhere in Europe... and that the effort devoted to such an enterprise would be likely to be ill-rewarded.'[77]

This view was perhaps not surprising when the position of the Army in 1940/41 is considered. Having been driven out of mainland Europe it required a huge programme of enlargement, restructuring, retraining and rearmament to produce an army equipped to fight its way back. Attention to the basics was essential. Airborne forces did not fall into this category and were regarded as an expensive luxury in terms of manpower and staff effort. At the end of the war experienced airborne officers believed that the endemic trouble with British airborne forces was that the Army had never really believed in them.[78] This was reflected in many cases by the quality of staff work connected with airborne development. This in turn provided a convenient hook for those in the Air Staff intent on censuring the airborne capability. The response to the General Staff paper 'The Value of Airborne Forces', was particularly uncompromising. 'The General Staff have succeeded in making a singularly unconvincing case for the value of airborne forces. The Paper would be a fair effort by a first year student at Camberley; but... as the supporting case for the expenditure of a substantial share of the national war effort at the expense of our Bomber offensive, it is a compliment to describe it as weak.' The Air Staff continued to pour scorn on the War Office effort, suggesting that the quotation of the parachute raid on Bruneval in February 1942 as an argument for maintaining a large airborne establishment demonstrated to what straits the General Staff were reduced in order to prove their case.[79] The Army had difficulty defending its case because it had neither the technical knowledge nor the political will to do so.

The latter point is illustrated by the comparative engagement with airborne development by the upper tiers of the War Office against the Air Ministry. At the very top Sinclair was a consistent if infrequent contributor to the dispute. Similar papers from Anthony Eden, Henry Margesson or Sir James Grigg, as successive Ministers for War, are practically non-existent. During the early part of the war the respective service chiefs were also unequally matched. Despite Dill's pre-war interest in airborne forces he now had far larger wolves much closer to the cabin door. He considered that he spent most of his time 'trying to prevent stupid things being done rather than in doing clever things!'[80] The CIGS found his task under Churchill intolerably taxing and he had little inclination to engage the Air Staff in a battle on a peripheral issue. On the other hand, Air Chief Marshal Sir Cyril Newall was fighting to validate and bring integrity to the RAF's central doctrine. Being Chairman of the COS Committee (until October 1940) enhanced his position, but crucially, he dictated the direction of the air effort because he was prepared to resign over the issue.[81]

The arrival of Portal as CAS in October 1940 did not alter the situation. A younger man than Dill, Portal had more energy and, as a staunch advocate of the bombing policy (having briefly headed Bomber Command), he had the strength and the will to oppose airborne forces development. He frequently pressed the Air Staff's position both in meetings and on paper. Dill did little to respond. The position became more balanced with the arrival of Brooke as CIGS in December 1941. Brooke was a firm promoter of airborne forces and a frequent contributor to their cause. Shortly after his arrival in office he issued instructions that development should be pushed to the utmost and given preferential treatment, and by the end of 1942 he was 'more convinced than ever that there is a great future for airborne forces.'[82] It was Brooke's unwillingness to compromise on the issue that forced it to be taken to the Prime Minister for arbitration. The fact that the decision fell in favour of the Air Staff, as has been shown, was the result of other government departments bolstering their position rather than because the General Staff argument was deficient.

Amidst this sea of resistance there were what appeared to be islands of cooperation rising above the rivalry. As has been described, the instinctive adoption of what would now be termed 'joint' working practices was a distant vision during the early part of Second World War. Special organisations were often required whenever two or more services were thrust together as part of the war effort. Airborne forces were no exception and the method for handling development and operations was laid down at an early stage.

> The development of air-borne forces is partly the responsibility of the War Office, Air Ministry, and DCO, but the Army Co-operation Command has been created for the primary purpose of developing all forms of air co-operation with the Army. It is therefore the responsibility of the AOC-in-C Army Co-operation Command to advise the Air Ministry on the tactical and technical air requirements for the development and organisation of air units and air training for air-borne forces in the British Isles. All development projects should therefore be referred to Headquarters Army Co-operation Command, policy matters being referred through the Directorate of Military Co-operation.

The same paper also stated that there might be an immediate requirement for airborne operations, probably on a relatively minor scale, and that DCO should be prepared to plan such operations.'[83]

Responsibility was therefore split during the early phase of development between Directorate Military Command (DMC) through Army Cooperation Command (AC Comd), for training and development, and DCO for operations. Although AC Comd's remit had only referred to the air requirements and air training, in reality the boundary was wider. The CLE came under their command and therefore, as will be seen later, the development of tactics and equipment also came under the responsibility of AC Comd and ultimately DMC. On the whole AC Comd did drive the development of British airborne forces between 1940 and 1942, mainly due to the fact that it was the only organisation in a position to do so. However, like many other 'joint' organisations of its time it was frequently hamstrung by the mistrust of both the services it was attempting to bring together. The Air Ministry was worried that the entire Army cooperation organisation might 'go native', and attempted to wrest decisions concerning support to the Army back to the centre. Slessor again was the detractor. 'It is important that dealings with them [the Army] over this [air requirements] should be done through the proper channels, and the proper channel in this case is me [Director of Plans].'[84] DMC must have been acutely aware of his weak position within the Air Ministry. The requirements of AOC-in-C AC Comd, 'the Cinderella of the Air Staff',[85] did not come close to the priority of those of Bomber and Fighter Command, and DMC did not hold the rank to compete on an equal footing with other departments such as the Director of Plans and the Director of Training. In addition Air Commodore R.V. Goddard during his tenure as DMC came under fire from the War Office as he was perceived as resisting the Army's requirements. In fact he was doing his best to ensure that the development of airborne forces was based on firm principles in order to build validity. This did not prevent him coming under sustained attack from VCIGS, Major General Archibald Nye. 'To hell with principles – give me the problem.... We are faced in fact with a practical problem which demands practical steps to be taken to meet it and a discussion on abstract principles seems to me will not get us anywhere.'[86] Goddard did his best to point out that practical steps were useless if they bore little relation to what was required, but the over-arching dichotomy made his position difficult.

The situation improved to a degree with a reorganisation within the War Office. For the first two years of airborne development the department charged with providing advice concerning the Army facet to DMC and AC Comd was part of the Director of Staff Duties (DSD). Lieutenant Colonel J. Stephenson of SD4 appears to have been the War Office's resident spokesman concerning airborne forces. He was industrious in trying to assist development but lacked practical experience. His Air Ministry counterparts usually vastly outnumbered him at meetings. When other General Staff officers did attend they frequently outranked him and then outflanked his well-intentioned efforts. Stephenson must have been exasperated during the meeting in which VCIGS announced that any effort devoted to airborne development would be ill rewarded. However, in June 1942 the War Office created a dedicated Air Directorate although it still fell under DSD. The new directorate was formed to sponsor the affairs of airborne forces and to act as a special link between the War Office and Air Ministry on all air matters. Brigadier Richard Gale was chosen to head the branch. Gale was already an experienced airborne officer having commanded 1 Parachute Brigade. His staff consisted of other, equally qualified, airborne experts such as Lieutenant Colonel Gerald Lathbury, many of who went on to distinguished airborne command during operations. A single point of contact had been created, with authority born out of experience, which could deal with the often conflicting requirements of the Air Ministry, War Office, the Airborne Division and the AFE. The Air Directorate provided knowledge and continuity within which policy could be developed, and relieved the Airborne Division Headquarters of many of its staff duties.[87] Gale was able to forge ahead with policy and improve links with other departments so that when the order was given for rapid expansion in order to achieve the requirements of HUSKY and ROUNDUP in early 1943, the Air Directorate was able to respond swiftly and effectively. Notwithstanding this the Air Directorate was still essentially a single-service organisation within the War Office. While it did good work in coordinating the War Office requirements with the Air Ministry it still had no power of compulsion over the latter and so the Air Staff could still resist or ignore the requirements of the airborne capability if it felt inclined to do so.

The second 'joint' organisation involved with airborne development was the DCO, 'an uneasy and unloved organisation – neither flesh, nor fish, nor fowl and distasteful to lovers of all three.'[88] While British airborne forces were still in the early stages of development the DCO was charged with identifying and planning those combined operations in which parachutists or glider-borne forces might be employed. In the language of the Second World War all airborne operations were considered 'combined' as they required the resources of at least two of the services to be executed. Combined operations was another 'Cinderella' concept at the outbreak of the war. Between the wars the requirement for combined training was recognised and the Inter-Services Training and Development Centre (ISTDC) had been established. However, under-funded, the ISTDC was briefly disbanded as war broke out.[89] Churchill established Combined Operations Command (COC) in July 1940. Lieutenant General Sir Alan Bourne of the Royal Marines (RM) was appointed the first Commander of Raiding Operations but was soon removed following a dispute over strategy with Churchill and replaced by Admiral of the Fleet Sir Roger Keyes in the post of DCO. Combined Operations began in an atmosphere of controversy and acrimony. They were disliked and mistrusted by all three established services, and came in for special loathing from the Admiralty.[90] Keyes did not assist COC's cause. He possessed drive and leadership in abundance, but he viewed his position as *supra* to rather than *intra* the services.[91] The COS Committee did not share Keyes's assessment of the position of DCO nor his enthusiasm for raiding operations and reached the stage where they were outwardly dismissive of proposed projects that bore his signature. In return Keyes considered the COS Committee 'the greatest cowards I have ever met.'[92] This lack of mutual trust manifested itself as a lack of willing cooperation, the antithesis of what COC had been established to achieve. This can be seen in Slessor's letter to Squadron Leader Louis Strange of the CLS. 'What I am really writing to you personally and privately about is to ask you very earnestly not to go talking to Roger Keyes… we must tread very warily with this DCO party.'[93]

Keyes was sixty-eight years old in 1941 and many of his staff were equally elderly. Their enthusiasm outstripped their endurance. Only one airborne operation was executed during Keyes' period as DCO. Operation COLOSSUS was Britain's first parachute raid, designed to destroy the Tragino aqueduct in Italy in February 1941. COLOSSUS will be discussed in more detail later; suffice to say at this stage the intelligence it was planned on was poor, as was the escape plan that relied on the cooperation of the Royal Navy. The result was that all thirty-five paratroopers that took part were captured. Keyes began to lose Churchill's confidence and in August 1941 Commodore Lord Louis Mountbatten joined COC as adviser to DCO. The writing was on the wall, and on 10 October, Mountbatten, aged only forty-one, took over as DCO. Mountbatten provided the political sense that Keyes lacked.[94] He took over 'a formidable inheritance of feuds' and in order to give him the required authority Churchill promoted Mountbatten to Vice-Admiral and gave him the equivalent rank in the other two services and a seat on the COS Committee. Mountbatten could now act decisively, 'There were still a lot of people about who saw nothing virtuous in combining the work of the three services. They were no use to me, but had been a pain in the neck in the past. Now, not only did I not have to use them, I could over-rule them. And did so.'[95] Mountbatten forged a far better relationship with the service chiefs and, as will be seen, was able to cut through resistance by being in a position to state his case personally at the COS Committee. A year after COLOSSUS, Operation BITING, Britain's second airborne raid, was launched to recover parts of a German radar from Bruneval on the coast of France. The operation was a success on all counts and the stock of British airborne forces and COC rose accordingly. Throughout the remainder of the war the relationship between DCO and COC and airborne forces continued to be healthy. Major General R.E. Laycock replaced Mountbatten in the autumn of 1943, and from HUSKY onwards, major airborne operations became part of the mainstream planning process. Airborne formations became subordinate to conventional operational formations as had been prescribed in the JPS's original concept for an invasion corps and DCO only retained a role overseeing minor special forces airborne operations.

During the first half of the war the DCO and the DMC within the Air Ministry suffered from the same difficulties that they had been established to overcome, namely a lack of cooperation between the services. They were subject to the same prejudices and friction that existed between the Army and RAF in microcosm. Those overarching problems were amplified in the case of the development of airborne forces by an imprecise and non-prescriptive, politically conceived requirement. Not until the planning for HUSKY did the requirement fully crystallise so that development could be accelerated and restrictions overcome, just as the War Office had predicted would be the case more than two years earlier.[96]

The Influence of other Government Departments

In addition to overcoming resistance caused by internal friction, the War Office and Air Ministry also had to deal with pressure from other government departments. There are examples of the requirements of other ministers coming into conflict with the needs of the service departments: Ernest Bevin's harassment of the War Cabinet for extra manpower for his Department of Labour is but one.[97] There were, however, departments that directly influenced the development of Britain's airborne capability, in particular MAP and the India Office. These departments brought pressure to bear (principally on the Air Ministry) resulting in an observable effect on the evolutionary path of British airborne forces.

Following the First World War the British aircraft industry had been allowed to slip slowly into a parlous state. Insufficient funding left the RAF equipped well below its own predicted first line requirements in terms of aircraft, and by the mid-1930s the Air Staff had some justification for believing that their service was being starved out.[98] Britain's best fighter and bomber aircraft lacked range, speed and payload. British aircraft firms were forced to seek alternative manufacturing business in order to stay solvent.[99] In 1935 the output of Britain's civilian aircraft industry was worth just £13,000,000, less than three per cent of national engineering output.[100] Despite this, when rearmament began the Air Ministry considered that it had in place plans which, once activated by government decision, could produce an air force second to none. These plans were carefully designed to mature in 1942.[101] In 1938 the Air Ministry had taken positive steps away from the methods of peacetime production by splitting the branch of the Air Member for Supply and Research into more focussed organisations under an Air Member for Supply and a separate Air Member for Research and Development.[102] These confident actions did not impress politicians who were concerned that on the outbreak of war Britain's monthly aircraft production would, with luck, not be much more than one half that of Germany, and was likely to be a good deal less unless drastic action was taken.[103] In order to counter the poor position of Britain's armed forces in terms of equipment, and to harness and control the country's industry, the formation of a Ministry of Supply was considered. There was an opinion that suggested that to be effective such a Ministry would have to include the supply departments of all three services.[104] The arguments for and against a Ministry of Supply were frequently debated in the press and the commons in the months prior to the outbreak of war. When Chamberlain announced on 29 March 1939 that the Territorial Army would be brought up to war establishment and then doubled he stated that, 'this important decision would involve a number of consequential decisions in order to provide *inter alia* for the necessary increases in equipment and reserves.'[105] The bill for the formation of a Ministry of Supply was announced in the House of Commons three weeks later. However, because the decision had been linked to the expansion of the Army, the new Ministry's functions were limited to the supply of Army needs only. Aircraft production, as predicted, was still woefully inadequate at the outbreak of war and continued to be so into 1940. Churchill had been one of those who had spoken out in support of an all-encompassing Ministry of Supply throughout the early period of rearmament.[106] On his appointment as Prime Minister he took swift and decisive action to rectify the poor state of aircraft production. He appointed a Minister of Aircraft Production on 17 May 1940 whose powers were confirmed in a bill of 20 May 1940. The creation of this new ministry reflected the urgency that the government now attached to the output of aircraft.[107]

The man appointed as Minister was Lord Beaverbrook, a personal friend of Churchill for nearly thirty years. Others did not share the Prime Minister's enthusiasm for the Canadian newspaper magnate. The General Staff's opinion of him appears to have been divided. Ismay and Major General Leslie Hollis clearly admired the results he achieved although the latter did note that he rode roughshod over everyone in order to achieve them and that his approach to business did not endear him to others.[108] Lieutenant General Sir Henry Pownall was not so generous, describing Beaverbrook as 'a totally unscrupulous cad of the Yellow Press' and considered that 'his strategical [sic] judgement is of course quite appalling... I doubt if his political judgement is any better.'[109] However it was Beaverbrook's relationship with the Air Ministry that had the most telling consequences. Shortly after the formation of MAP Sinclair informed Churchill that he struggled to keep on good terms with Beaverbrook but that he could be relied upon to continue struggling.[110] Their relationship was predictably dysfunctional, after all Beaverbrook had just removed from Sinclair the means to meet his own requirements in terms of aircraft and ammunition. Not only had Sinclair lost a crucial sector of his ministry but MAP was after more.

MAP under Beaverbrook had predatory instincts. With each concession that the Air Ministry made Beaverbrook demanded more. 'Let me say that your [RAF] problems of shortage of aircraft are being solved by this Ministry and by no other agency' he wrote to Newall, 'I should think that this service might be recognised by placing at our disposal forthwith all those portions of the Air Ministry now engaged on the production of aircraft. That is all I ask. And I cannot understand why it isn't given to me at once.'[111] Although on the surface reasonable, the Air Staff regarded MAP's requests for expansion as dangerous, not just to the autonomy of the Air Ministry but also to the war effort itself. In particular Beaverbrook's insistence that the department of the Deputy Director General of Equipment (DGE) and the whole of Maintenance Command be transferred to his authority caused intense rancour. He demanded the shift of these departments in June 1940.[112]

DGE had been reorganised in April 1940 in order to streamline aircraft production by dividing the department into three.[113] The first department was charged with provision of complete aircraft along with spare parts and equipment necessary for handling and maintenance. The second had responsibility for procuring ancillary equipment including ammunition, explosives and supplies. But it was the transfer of the third that caused so much acrimony. This department was responsible for equipment planning, administration and movement. To the Air Staff the danger was obvious.

> Clearly it is a function of the Air Ministry to formulate the Air Force requirements in equipment and to set the task to the Production Staff of meeting these requirements. This has always been accepted as a sound principle. It is fantastic to expect the Air Council to transfer to any other Department the staff that is responsible for formulating its equipment requirements and for watching the progress of these demands through the Finance, Contracts, Development, Production and Inspection stages and for constantly pressing that the requirements are met at the correct times and in the correct quantities.[114]

Without being able to set its own requirements the Air Ministry could not be held accountable to Parliament for the efficiency and effectiveness of the RAF. There was also the threat of a tendency for the requirements of 'difficult' items of equipment to be reduced so as to meet production capacity.[115]

The question of the transfer of Maintenance Command was opposed with equal vehemence. 'The second and far more serious danger is that the Air Ministry should surrender to any other authority the staffs and the Units which control the stocks of equipment and the reserves of all kinds which form part of the fighting equipment of the Royal Air Force. Such a transfer would take out of the hands of the Air Council a function which is vital to the control of aircraft operations.'[116] The RAF would in effect have lost the ability to decide where the main effort for repair of in-service aircraft should be directed and where reserves of ready aircraft should be held in anticipation of the threat from Germany. The Air Staff could not tolerate this. 'It [The Air Ministry] must have full power to decide where the various reserves are kept because this is inseparably linked with Air Force organisation and administration. A few days after Lord Beaverbrook obtained responsibility for the manufacture of aircraft he was attempting to dictate what use should or should not be made of the aircraft once delivered.'[117]

Beaverbrook's demands for control of storage units, the ferry pilot pool and other departments had all been met but enough was enough. Sinclair thanked him for his request to take over DGE and Maintenance Command but concluded 'This is a proposal to which I could not possibly agree and

I hope you will not feel it necessary to pursue it.'[118] Newall went a step further and wrote direct to Churchill with reference to the Air Ministry's previous sacrifices to MAP. 'I have already gone a great deal further than is sound from the point of view of functioning and fighting the Royal Air Force in order to meet the views of the MAP and I hope you will see it possible to persuade Lord Beaverbrook to accept the arrangements which I propose as a final statement which I can assure you, we will do our utmost to operate in a spirit of mutual confidence.'[119]

On 1 August 1940 an agreement between MAP and the Air Ministry was written. While MAP did gain control of Maintenance Command the distribution and allotment of aircraft was specifically retained by the Air Ministry. MAP was also explicitly denied the transference of any part of DGE. This particular spat may not have had direct impact on the development of airborne forces but it is indicative of the acrimonious relationship between the two ministries. That discordant state existed almost immediately on the formation of MAP and influenced all discussions with the Air Ministry, including those that would directly affect airborne forces, specifically the question of production priorities.

Beaverbrook clearly had thoughts on where production priority should be directed, or not directed, before he ever became MAP. He was of the opinion in October 1939 that 'The bomber is a disappointment in war. It cannot stand up to the fighter, and it is beaten by the anti-aircraft gun.'[120] He repeated his assertion in March 1940: 'If mass bombing attacks were made on London the attacking force would be fatally damaged before they could achieve any real success.'[121] Beaverbrook was prejudiced against the bomber and therefore against the RAF's core doctrine. Either deliberately or subconsciously he allowed this bias to influence MAP's priorities for aircraft production when he first took office.

Beaverbrook first communicated his revised programme for production to Sinclair on 11 June 1940. However before the Air Staff had the opportunity to respond to the document Beaverbrook had written to Churchill complaining that the Air Ministry was obstructing his endeavours. He listed all the cases, both real and perceived, in which the Air Staff had frustrated him. He then attempted to rescind the letter.[122] This method of bluff was frequently practised by Beaverbrook with Churchill and is reflected in his many threats to resign. The Air Ministry had much to consider in the revised production programme. The requirements of the bomber offensive were woefully catered for. The figures for the Vickers Wellington, the only modern bomber on the programme, indicated a production rate of 92 aircraft per month in June 1940 rising to 177 in June 1941.[123] This was compared with 300 Hawker Hurricanes and 135 Supermarine Spitfires in June 1940, rising to 410 and 245 respectively in June 1941. The only other bombers on the programme were the aging Armstrong-Whitworth Whitley, which was to be the mainstay of parachute training and operations during early airborne forces development, and the Armstrong-Whitworth Albermarle, which did not begin production in reasonable numbers until January 1941. The Short Stirling, Avro Manchester and Handley-Page Halifax had been ordered in 1937 to begin production in 1940 in order to attain a target of 3,500 deliveries by April 1942. None of these four-engine types appeared anywhere on Beaverbrook's programme and the original target slipped by a year as a result.[124] The ratio of fighter production to bomber production stood at just over three to one in June 1940 rising to nearly four to one in June 1941.

Understandably the Air Staff's response was not favourable, declaring that the offensive side of the RAF should not be unduly curtailed in favour of the defensive.[125] They were not happy that the acceleration in fighter production was not matched by similar development with regard to bombers. Although raw production figures could never be considered the sole measure of effectiveness, it was pointed out that the Air Ministry's pre-war production plans had aimed at a fighter to bomber ratio of less than two to one and now that airbases in France and the Low Countries had been denied the requirement for long range bombers to strike at Germany had increased further since those plans had been made.[126] MAP countered the Air Staff's alarm by explaining that the programme was the best that could be achieved with the available resources, in particular raw materials.[127] There is some substance to this statement as multi-engine aircraft used more high-tolerance materials than single engine fighters.

The production programme was issued and debated in the days leading up to Churchill's minute calling for the formation of airborne forces. MAP's priorities would have far reaching consequences for airborne development, as the bomber would be the only means of airborne delivery for the first two years of their development. Although they did not form the link with the bomber at this stage, the Air Staff were quick to grasp the effect that Beaverbrook's programme would have for airborne forces. Transport aircraft were practically non-existent in MAP's plan despite their importance being recognised. 'The [De Havilland] Hertfordshire is a transport aircraft on which, as a class, there have been heavy demands in recent months. Such aircraft have carried out work of a most important nature in the past, and there are bound to be even greater demands in the future, including possibly the conveyance of parachute troops in offensive operations.'[128] The Hertfordshire did not appear on the programme at all but the plea failed to secure any amendment. British transport aircraft numbers never reached an adequate level to effectively support airborne forces as will be described in the following chapter. Beaverbrook's insistence on accelerating fighter production did contribute to victory in the Battle of Britain. However, whether this was due to his foresight, considering his 'appalling strategical [sic] judgement', or was simply a consequence of his prejudice for the fighter over the bomber can be debated.

His decision did have enduring consequences for Britain's offensive aspirations. Air Chief Marshal Sir Philip Joubert observed, 'Lord Beaverbrook, to put it bluntly, played hell with the war policy of the RAF. But he most certainly produced the aircraft that won the Battle of Britain. What he did in the summer of 1940 set back the winning of the air war over Germany by many months.'[129]

What he did to the air war over Germany he also did to airborne development. Airborne forces relied on bomber aircraft throughout the war but almost exclusively until late 1942. During this period operations and some areas of training were often frustrated by a lack of aircraft as will be seen later. As previously examined the Air Staff did resist airborne development in the initial stages; however, as Portal pointed out, if somewhat disingenuously, 'the disappointments experienced by Army Cooperation Command are due, not to a lack of good will on the part of the Air Ministry, but to hard realities of the aircraft supply position over an acutely difficult period.'[130]

Beaverbrook attempted to resign more than once during his tenure as MAP, more usually through a fit of pique or in an attempt to bring the perceived machinations of the Air Ministry to the attention of the Prime Minister rather than because of any sincere desire to step down from the appointment. However in spring 1941 he did resolve to go, and at the beginning of May, Churchill moved him into the post of Minister of State supervising the three supply ministries and acting as referee in priority questions. Beaverbrook could not resist one final parting shot at Sinclair. 'It would be better if the Air Ministry devoted more time to operations and less attention to the affairs of the Ministry of Aircraft Production, which seems to have served their needs so far.'[131] In reality he had served the needs of the nation by sacrificing the wants of the Air Ministry. While defending Britain he badly damaged the relationship between the two ministries, which had an effect on all negotiations from May 1940 until May 1941 and beyond. This in turn had a commensurate effect on airborne development during this period. As will be seen, this became manifest not only in the lack of suitable bomber aircraft for airborne training and operations but also during the production of gliders and early attempts to secure aircraft through production in the United States.

The pressure applied by MAP was principally physical in nature and had a negative effect on airborne development through the deprivation of suitable aircraft. The pressure applied by the India Office is more difficult to define. It was not physical but doctrinal in nature and could even be perceived as positive in its effect on the evolution of British airborne forces. The chief protagonists were the Lord Linlithgow, the Viceroy of India until October 1943, Leo Amery, Secretary of State for India from May 1940 until the end of the war and Wavell as Commander in Chief (C-in-C) India from June 1941 and then as Viceroy on succeeding Linlithgow. Linlithgow made a proclamation of war on behalf of India on 3 September 1939. He did so without consulting Indian leaders and although most of them were happy to support the defence of the British Empire against German aggression there was an undercurrent of unease that this initiative did nothing to settle. Some saw the promise of support to the Empire during the war as a valuable bargaining chip for future independence. Others had foreseen more direct benefits and believed that Japanese invasion would leave Indians in charge of their own destiny. The country was ripe for revolution. Should that revolution break out then maintenance of control would become difficult in such a vast country.

Very early during the process of development it had been recognised that there appeared to be greater opportunities for operations to be carried out by airborne forces in the Middle East than Europe.[132] The factors behind this reasoning included topography, meteorology and the lower density of enemy forces in the Middle East against those in Europe. The JPS's paper detailing future plans and basic requirements called for two invasion corps to be formed, one in Britain and one in the Middle East. Each identical invasion corps included an airborne element.[133] When the COS Committee ratified this requirement and agreed the formation of an airborne brigade in India it noted that 'Quite apart from any use for a brigade of this nature overseas the Government of India consider that it would be most valuable for employment for the Defence of India and for Internal Security.'[134] Here lies the reason why India became so enamoured with the possibilities of airborne forces and why pressure was brought to bear on the War Office, Air Ministry and the Prime Minister.

In the spring of 1941, as part of the operations by 10th Indian Division to re-establish British control in Iraq, 1 Battalion, the King's Own Royal Regiment with support was flown at short notice onto the airfield at Habbaniya in order to break its siege by Iraqi forces. The operation was successful despite losses in aircraft during the approach into Habbaniya.[135] The possibilities of using an airborne capability to swiftly counter any uprising within the Empire were not lost on observers in India. Linlithgow came to adopt the concept as his own declaring 'I have personally nursed our infant airborne forces with tender care because I am convinced that their contribution is going to be essential.'[136] However it was Amery who led the Viceroy along this path.

Like Beaverbrook, Amery was a personal friend of Churchill, having been at Harrow with the Prime Minister. With such a close relationship Amery was disappointed when only offered the India Office by the Prime Minister in May 1940. However Amery had a long history in colonial affairs and crucially he and Churchill had overlapped by six weeks in the Colonial Office in 1921. Amery had taken this opportunity as Churchill's subordinate to follow through 'pet' projects and policies. In particular he finished the campaign against the 'Mad Mullah' in Somaliland, delivering the final blow using a squadron of twelve aircraft from Egypt that bombed the uprising into submission.¹³⁷ Amery was therefore already well aware of the advantages of the efficient use of air power in an imperial internal security role and he knew that Churchill was also conscious of the potential.

Amery, akin to Beaverbrook, aroused mixed emotions among the military community. Pownall was forthright once more describing Amery as 'a hopeless and stupid little creature.'¹³⁸ The COS Committee approved the formation of an Indian airborne brigade in June 1941 subject to the caveat that at least one of the three constituent battalions should be British.¹³⁹ Progress was initially slow, as it was in Britain, but pressure on the military establishment from Amery to speed up development began in a polite fashion. At first he queried Dill: 'I wonder whether anything more is being done to consider Wavell's request for a larger measure of airborne preparation?'¹⁴⁰ In reply CIGS pointed out that the rate of development was largely reliant on the availability of aircraft and gliders and that it would be difficult to allot specialised types of aircraft specifically for the use of an airborne formation.¹⁴¹ Concurrent to his approach to Dill, Amery also began to broach the subject with the Prime Minister. Again referring to Wavell's request and to the success at Habbaniya, Amery presented an extremely well informed case for airborne expansion in India and expressed his concerns over lack of progress. 'I gathered from a memorandum by Pownall which Dill showed me that [airborne expansion] is regarded as entirely out of the question as there will never be enough bombers available to tow gliders.'¹⁴² Amery thought that this was a negative conclusion; however, thanks to Dill, he now had his own idea of where airborne development was being obstructed. Despite CIGS pointing out that the lack of suitable aircraft was the fault of MAP, Amery aimed his frustration at the Air Ministry. In a second letter to Churchill he highlighted once again the success at Habbaniyah and postulated that General Sir Claude Auchinleck might have been able to stop Rommel on the Tripoli road if he had available an airborne brigade to fly forward. Amery also predicted the use of an airborne capability in seizing and re-establishing control over areas of the Mediterranean and in regaining territory in Burma. Airborne forces stationed in the Middle East or, more precisely, India could serve both these objectives. 'I still believe that the only way to get the thing on an adequate scale is to insist on having it, on whatever scale you decide on, entirely separate from the Air Force.'¹⁴³ The Prime Minister was obviously impressed by the argument and added a comment to the bottom of the letter for the attention of CAS; 'There ought to be an extra Airborne Division – or better its equivalent in Bde groups – raised in India as soon as possible. Pray consider this and let me have some proposals.'¹⁴⁴

Portal's response was understandably robust and well rehearsed as he was already under similar pressure directly from the War Office. He stated once again that the rate of development was a direct consequence of the availability of aircraft, a state of affairs that at this stage of the war could not be accelerated. 'It is an entire fallacy to suppose that there is an unused reserve of productive capacity which could be turned on to making transport aircraft and special engines for them. So far as this country is concerned such capacity could only be found by a cut in the deliveries of some other type of aircraft.'[145] It was a familiar and perfectly logical argument but Amery was not satisfied. He summed up the position as he saw it: 'Lack of suitable aircraft and statichutes [parachutes] is delaying training of our parachute brigade, and [the] picked force which might be employed elsewhere is being wasted at present,' and it was 'useless to continue effort to provide these forces in the present atmosphere of lethargy and indifference.'[146] It was at this point that Linlithgow adopted airborne forces as his personal 'infant'. Amery appealed to his ally Churchill once more. 'If at this moment we could transport and supply even a brigade from India to Burma it might make a big difference.... If so, then, whatever the demands of the Air Force for other purposes, there should be no question as to the necessity of sparing some proportion, however small, of the total for training and building up air transport.'[147] The logic was inescapable but it was not drawing the desired response. The Air Ministry was only just beginning to receive significant numbers of modern bomber aircraft and it was not intentionally going to dilute its central doctrine at this stage of the war by allocating bombers to airborne forces or switching production to transport aircraft. Portal was already defending this corner against Brooke, and while Amery failed to coordinate his campaign with the War Office he was easily fended off.

Linlithgow, having been convinced of the advantages of airborne forces, was now becoming frustrated with the lack of progress. He urged Amery to continue to apply pressure to Churchill. 'I have done my utmost personally to encourage the pioneers of this branch of warfare from the very start because I felt that we were sure to need them in this theatre. It is now clear that we shall want them badly.'[148] The Viceroy was correct in his assumption; the arrest of Congress leaders on the evidence of an impending revolt was only two months away. It helped to validate Amery's theory that it might be best 'to go back more to the spirit of Mutiny days and revive British Rule in its most direct and, if necessary, ruthless form.'[149] During the riots that followed the Congress arrests Indian parachute units were employed in Delhi, and although their airborne capability was not needed their utility in the internal security role was demonstrated. Despite minor parachute operations being mounted in the Sind Desert and in the Myitkyina area of Burma in July and August 1942 even Amery was by mid-1942 forced to admit that the chances of building up an airborne division in India appeared hopeless.[150]

In fact the situation for British airborne forces was beginning to look hopeless on all fronts. In a few months CIGS and CAS would take their disagreement on the subject to Churchill for arbitration. By the end of the year the decision would be made to reduce the British airborne organisation to two brigades and any possibility of expansion in India was lost. It was not until the potential of the airborne capability had been demonstrated during TORCH and HUSKY and the means to meet the JPS's future requirements were available that development in India was once again considered. In September 1943 Browning visited India on the orders of Nye to report on the situation of airborne forces. The result of Browning's report was rapid expansion and the formation of 44th Indian Airborne Division during 1944.

Amery was a persistent champion of airborne forces in India during 1941 and 1942. He formulated a persuasive argument for the employment of the new capability. His original vision did not materialise, as airborne forces were not required to use their strategic mobility for internal security duties within India during the war, although the role was fulfilled by 2 Independent Parachute Brigade in Greece in 1944. In fact Amery's vision could not begin to be fulfilled until the strategic situation in 1944 allowed the diversion of suitable aircraft from Europe to India. This came too late for British airborne forces in India to have a major role in the war in the Far East. Only one major airborne operation was conducted when a composite Gurkha Parachute Battalion landed at Elephant Point during Operation DRACULA, the assault on Rangoon in May 1945. Against Amery were his constant attempts to use Churchill to influence the situation that put the Air Ministry further on the defensive and increased their resistance to the airborne concept. Had Amery and Wavell coordinated their efforts with the War Office then resistance might have been more difficult. As is usual when inter-departmental conflicts arise the result was the diversion of considerable staff effort that could have been more profitably expended elsewhere. The diversion of aircraft, manpower and equipment from Europe to India was not in the end significant. There was, however, one unforeseen effect of the expansion of India's airborne forces in 1944: the European theatre lost the man widely considered at the time to be Britain's most proficient airborne commander. The experienced Major General Ernest Down handed over command of 1st Airborne Division to Major General Robert 'Roy' Urquhart in January 1944 in order to take over the fledgling 44th Indian Airborne Division, and as a result he never saw action as an airborne commander. The repercussions of this change in command will be explored later.

The Political Environment as a factor in Development
The political environment was critical to the development of British airborne forces, particularly between June 1940 and January 1943 and

during the SYMBOL conference. Prior to the Casablanca conference there was no obvious strategic or operational imperative for the employment of airborne forces. Britain was locked in a defensive battle on several fronts and an airborne capability appeared to be an expensive luxury. The work of the JPS on the airborne concept was still largely theoretical. Without any firm model or requirement on which to base planning the ministries felt justified in doing the minimum to satisfy the requirement. Without a clear and imminent case for the employment of airborne forces the resistance of the Air Staff and the indifference of the General Staff can be understood if not justified. A strong vision of the future utility of airborne forces and the authority and will to promote the concept was necessary to ensure progress. This fell to Churchill and the inception of the capability was a result of his *diktat* as opposed to any identified military requirement. However once he committed the Army and RAF to airborne development Churchill failed to act effectively as a 'progress chaser' and there is little evidence that 'he removed bottlenecks and hastened growth.'[151] Nor was he a flawless omnipotence, 'picking up all the threads and giving them coherent shape and form.' He certainly did not 'continue to pester, nag and bite.'[152] The more balanced assessment of Churchill as a uniquely able but flawed leader most aptly expresses his actions in this situation. In the case of airborne development that unique ability was demonstrated by his initial vision but his flaws did not manifest themselves as 'blunders and hasty, impetuous decisions,'[153] but rather as inconsistent interest and the inability to make a timely and effective decision when called upon to do so. He did show interest in airborne forces throughout the first part of the war but in an irregular and highly punctuated manner with no real follow-up to his occasional visits and questions. These aberrations can be excused when they are placed beside the Prime Minister's fantastic workload and the list of significant strategic priorities that were competing for his time and effort, and the lack of consistent advice that he was receiving from his staff and advisors.

It has been claimed that, 'The general effectiveness of the British political–military response to the Second World War rested on one of the most efficient decision-making systems of the major powers involved in the conflict.'[154] This still stands but it should be recognised that Churchill's minute of 22 June 1940, vague and open to misinterpretation, was made more difficult to implement by the fissures in the system of defence coordination and planning. Below the Minister of Defence there was little synchronisation between the services. Therefore fundamental disagreements over the structure, organisation and employment of future airborne forces became impossible to resolve at COS level. Concurrently the JPS were working on the long-term concept and doctrine of the capability and how this would contribute to the future direction of the war, in particular the invasion of mainland Europe. DCO was meanwhile responsible for the employment of Britain's fledgling airborne forces and was therefore learning valuable lessons, which could inform future operations. AC Command, operating within the RAF's stove-piped infrastructure, was independently developing training and tactics. There was little formal constructive interaction between these organisations and therefore development was not being managed in a holistic manner. When the Prime Minister did intercede his observations and criticisms were generally followed by a surge of staff activity but little practical action. When the organisations did coincide on the subject the result was often negative, primarily due to the pervasive scarcity of resources. Both these effects were due to the failure to establish a dedicated, cross-service committee or organisation to specifically deal with the complex requirements of airborne development. The overall result was that the evolution of British airborne forces was extremely slow up until January 1943.

Following SYMBOL and the identification of an assault on Sicily as the next allied strategic objective, the need for an airborne force to assist the amphibious landing and prevent a repetition of the Dieppe debacle became obvious. From that moment forward the General Staff and, to less enthusiastic extent, the Air Staff, committed themselves to the provision of an airborne capability. With a clear military requirement the need for political direction receded and strong policy was more forthcoming. Even long held inter-service rivalries were to some extent overcome. Political compulsion and bureaucratic impediment therefore only contributed one dimension to the development of airborne forces during the Second World War. This requires balancing against the impact of resource constraints, both equipment and personnel, and the effect of the strategic purpose on the development of concepts and doctrine in order to create a full conclusion tracing the development of British Airborne Forces.

CHAPTER THREE

Equipment and Technology

Personal and Support Equipment

To be effective a new military capability requires coherent development across all its constituent lines of development. Notwithstanding this it is not unusual for one particular aspect to be identified as being fundamental to its entirety. It is often this line of development that initially defines the capability and is the primary element around which the other lines are developed. This primary element of a new capability is often characterised by a novel piece of equipment or an innovative concept. The introduction of the British airborne capability during the Second World War was concept led. There was no single obvious piece of equipment associated with the new capability, rather an entire range had to be developed to ensure effective progress; from clothing and personal arms, through vehicles and support weapons to support aircraft and gliders. All would have to compete for priority in the struggle for scarce resources. Each would present differing degrees of technical and engineering complexity, and therefore progress across the equipment line of development was unlikely to be even. Much of the equipment would have to be developed through trial and error, as there was little experience in this area within the research community. Despite this, most of the equipment that paratroopers or airlanding soldiers required once they were on the ground was not very different from that used by any other infantryman of that era. This included clothing and personal weapons, some form of mobility and fire support if they were to survive in the face of the enemy, and communications in order to coordinate their effort. What was crucial was that this equipment could be delivered to the battlefield by the same means as the soldier. Likewise, engineers, medics and other supporting arms would have to adapt their equipment so that it could be successfully delivered from the air. The critical factor throughout the range of support equipment was weight and size, which had to be minimised wherever possible in order to maximise aircraft capacity.

When the CLS was formed at Ringway, technical trials, development and experimentation were carried out in an *ad hoc* manner with whatever limited resources were available at the time. It was not until the CLS was expanded and renamed the CLE on 19 September 1940 that a formal Technical Development Section (TDS) was established. However, the CLE was primarily an air-focused organisation and only one officer was committed to working on Army requirements. On 15 February 1942 the TDS became part of a new unit, the Airborne Forces Experimental Establishment (AFEE), and on 1 July it moved from Ringway to Sherburne-in-Elmet.[1] The AFEE had a greatly expanded establishment but still only made a single officer responsible for the development of paratroop equipment, albeit he did have a small team to assist with practical trials.[2] This level of commitment was clearly not going to be sufficient to develop the plethora of new, modified or adapted equipment that paratroops and airlanding soldiers would require. The War Office and the airborne units themselves would have to adopt some important principles and be receptive to other methods of acquisition if the airborne force was to be properly equipped.

In some cases the necessary modifications or adaptation of current equipment could be quickly identified and relatively simply accomplished. Even personal equipment and clothing had to be scrutinised for potential weight reductions. Greatcoats and ground sheets were discarded and new windproof, water-repelling smocks were produced to compensate. New, warm underclothing was developed. The standard British infantry helmet had the protruding rim removed, not only to save weight, but also to reduce the chance of it snagging in the aircraft or during landing. Rubber-soled boots were produced to reduce the shock of landing but these were later discontinued due to the shortage of rubber and a lack of durability.[3] Much of this clothing was based on German designs – so much so that the early British paratrooper was practically indistinguishable from his *Fallschirmjäger* counterpart.[4] The provision of specialist clothing did not create undue difficulties, as it did not present a high level of technical challenge. Nor did the scale of supply cause problems as civilian manufacturing capacity in the clothing industry had been drawn upon early in the war and had been developed with great rapidity.[5]

The Lee Enfield rifle and Bren light machine gun remained the standard section weapons. The Sten machine carbine was provided on a greater scale than in normal infantry units, at two per section, in order to provide some concentrated firepower from a lightweight weapon. On initial scrutiny, in view of the weapon's size and rate of fire, it might have been considered effective to issue more Stens per section. However, aside from the issues of range and accuracy when compared to the rifle, there were also logistic implications to issuing more Stens. Automatic weapons always consume more ammunition than single-shot weapons and this increase in expenditure has to be sustained. Rifles and Brens were issued to 353 personnel within a parachute battalion, which required first-line ammunition holdings of less than 60,000 rounds. The Sten was issued to 226 soldiers and yet required over 64,000 rounds of ammunition to be held at first-line.[6] Therefore any increase in the issue of Stens would result in more ammunition that had to be held within the battalion and delivered to the DZ. This may appear a minor point but these are the issues that had to be considered in order to balance firepower against the necessity to reduce aircraft loads. The 3-inch mortar remained the battalion's own integral means of providing indirect fire support in both parachute and airlanding units.[7] None of these weapons needed any major modifications and served their purpose with airborne forces throughout the war. In 1943 a detachment of the Army Operational Research Group (AORG) carried out observations on the equipment of an airborne battalion and concluded that, although minor equipment problems did exist, they were no different to those experienced by a normal infantry battalion.[8]

Paratroops and their personal and support equipment had to be delivered from their aircraft to the battlefield safely and effectively. The development of the parachute through its various types and marks has been described in detail elsewhere.[9] Despite occasional accidents resulting in fatalities the British parachute, particular the X-type (known initially as the 'statichute'), was remarkably successful. Notwithstanding this, the scale of the challenge in providing parachutes to the airborne force is worth further examination. Shortly after Churchill's minute of 22 June 1940 the Treasury reported that 10,000 parachutes were urgently required for 'highly secret purposes.' Later that year the MAP estimated that it would require 17,200 parachutes per month to equip airborne forces.[10] These figures caused consternation within the Treasury. Each 28-foot diameter X-type parachute required 68 square yards of silk, which would equate to a monthly consumption of over 1 million square yards. MAP's figures meant that the parachute requirement for the twelve months from September 1940 would account for 85 per cent of the national stocks of silk that had originally been calculated to be sufficient for two years. Moreover, parachutes were not the sole military consumers of silk. Charge bags for naval and military guns required approximately 400,000 square yards per month. Silk reserves were unlikely to improve as the attitude of Japan, Britain's major supplier, grew increasingly uncertain through 1941.[11] When Japan did enter the war the situation became critical.

Part of the problem was the expendable nature of the parachute. It was assumed correctly that parachutes used during operations would be subject to a 100per cent wastage rate and therefore an entire formation's parachutes had to be held in reserve for each projected operation. In addition to this a pool of parachutes was required to cover training and maintenance. This calculation indicated that an airborne brigade which would deploy 1,880 soldiers by parachute would be required to hold 5,260 X-type parachutes to enable training and maintenance and to provide an operational reserve.[12] Once these figures were extrapolated to cover the entire projected airborne organisation, resupply parachutes were added to the equation and then the total was multiplied to cover multiple operations. The War Office estimated that the total number required for the fifteen to eighteen month period from May 1942 equalled approximately 140,000 parachutes. The War Office requested the Air Ministry to consider stockpiling raw materials in order to ensure that production was not prejudiced but the state of the international silk market made this difficult.[13]

Late in 1942 shortages were having a significant effect on supply. By October only 17,687 parachutes had been delivered, well below the planned programme. The situation was not alleviated by an inability to obtain raw silk from the US as it was retaining indigenous production for its own requirements.[14] Silk production in India could also not be relied on. India's own airborne development programme was outstripping local supply. By mid-1944 Indian airborne forces had a projected deficit of 403,000 parachutes.[15] MAP attempted to mitigate the apparent shortfall by requesting that the US supply Britain with completed parachutes, and by investigating alternative materials suitable for manufacture. Various natural and man-made fibres were tested. Cotton was cleared for use for resupply parachutes, and a cotton and man-made fibre combined material known as Nylex or Ramex was cleared for use in X-type manufacture. Nylon was also tested late in 1942 but the results were inconsistent and it was not cleared for use at that time.[16] Nylon did not become universally available for parachute manufacture until 1944. With this mitigation in place and by reducing the overall requirement from 140,000 to 60,000 parachutes, the War Office was satisfied that its demands would be met in time for OVERLORD.

The most noteworthy fact concerning the mass supply of parachutes in Britain during the war is that at no point, apart from in India very early during development, does any shortage of parachutes appear to have impinged on training or operations. There are a number of explanations for this although there is no evidence to support one of them above the other. The situation may have been extremely well managed at the front line so that shortages were not apparent to the user. Alternatively it is possible that production became misaligned with consumption so that even when the manufacturing rate fell due to silk shortages output was still sufficient to meet the actual demand. It is also probable that the War Office over-estimated the requirement for parachutes, inadvertently or

Equipment and Technology

even deliberately in order to ensure that the actual requirement would be met.

It became clear relatively early in the course of development that a paratrooper could not drop while carrying a rifle, Sten or Bren as this greatly increased the chance of injury on landing.[17] The same was true of radios, mortars and other support equipment that a normal infantryman might expect to carry with him as he walked or rode into battle. A system was required to deliver this essential equipment to the battlefield as close as possible to the men who needed it. A number of containers were developed, capable of carrying weapons, radios and general stores.[18] These could be dispatched via the same fuselage aperture as the parachutist or from the bomb racks of support aircraft. The various types of container and their functions have been described in detail elsewhere but again a brief examination of the scale of supply is warranted.[19] A section of ten men required two containers to be dropped with them containing their personal weapons and immediate ammunition requirements. A 3-inch mortar section required four containers to carry the mortar, bombs and personal weapons of the men.[20] When the figures were aggregated and a scale for wastage and training was applied a parachute battalion required 274 containers of all types.[21] When viewed across the entire airborne force the War Office estimated the requirement for eighteen months added up to approximately 34,000 containers.[22] Although there was never apparently a shortage of containers, the quantities involved did generate an interesting debate.

The provision of aircraft for airborne forces was clearly an Air Ministry responsibility while the equipment to be used by the soldiers on the ground belonged to the War Office. Containers had to fit onto or into aircraft but also had to be capable of carrying the Army's equipment, hence the responsibility overlapped the two departments. The Air Ministry was positive that it bore the liability for the provision of all aircraft fittings including static lines, parachutes and containers, and that the War Office was responsible for stating what needed to be carried within the containers.[23] The War Office did not dispute this but wished it to be confirmed that the Air Ministry was also responsible for the storage of containers. Gale, as the commander of 1 Parachute Brigade cited the argument that the RAF accepted responsibility for airborne operations from embarkation until the troops landed on the DZ, therefore the RAF should hold the containers on their airfields until required.[24] This was a considerable undertaking as the average container was two metres long and nearly half a metre in diameter, and some were nearly double this size; the storage of 274 for a single battalion would take up substantial space. Unsurprisingly, the RAF resisted this initiative, suggesting that containers should be held by the airborne units that were going to use them.[25] A sensible compromise was reached and it was decreed that the Air Ministry would continue to be responsible for the research and provision of parachutes and containers and would also be responsible for the storage of parachutes. The War Office would be responsible for the storage of containers but the Air Staff agreed to examine the possibility of forming RAF holding and packing units to assist airborne units with this task.[26]

This relatively minor incident illustrates the schism that existed within the airborne equipment development domain. The Air Ministry was not interested in any airborne equipment that was not directly linked to aircraft. Until the AFEE was formed this attitude permeated through to the CLE, which essentially had an RAF chain of command. At the same time that Gale was arguing about the storage of containers the CLE's superior commander, AOC No. 70 Group RAF, reported, 'little thought or development work has yet been given to the special equipment required by the Airborne Brigade Group.'[27] At this point, late in 1941, British airborne forces had already been committed to one minor operation, were expecting to be committed imminently to others and were busy training for these contingencies. Equipment faults and shortfalls were often identified during training, and, with the CLE's apparent lack of progress in this area, it is unsurprising that the airborne troops themselves developed and adopted improvised solutions. Paratroops quickly realised that the container was not an ideal method of delivery for personal weapons. Lieutenant Colonel Frost, the commanding officer of 2 Battalion, the Parachute Regiment described the problem.

> The drill was that the weapons containers were dropped either in front of the parachutists, or in the middle of a stick of parachutists, or at the end. This meant that before the parachutist could be effective he had to find the weapon containers in the dark, sort them out and extract the weapons. We had found that this was extremely difficult in practice and took a lot of valuable time. Indeed some containers were never found at all.[28]

Frost gives his own battalion the credit for creating the solution to this problem from its own resources, although the official history acknowledges Major John Lander of 21 Independent Parachute Squadron as the pioneer. Which ever is the case the kitbag and valise certainly appear to be have been initially developed from within the front line rather than the development community. The kitbag was made of padded canvas and leather and was large enough to hold a Sten, radios and other small but bulky pieces of equipment. The valise was designed to hold a rifle or Bren. Both were attached to the paratrooper during embarkation and while in the aircraft. Once he had jumped a quick release was operated and the kitbag or valise was lowered on a rope approximately seven metres below the man. This meant that the weight of the equipment did not have an undue effect on the paratrooper during landing but his equipment was easily retrieved on the end of a rope. This system was successfully operated from mid-1942 onwards, although containers continued in service for resupply and for larger items of equipment that were not required immediately on the DZ.[29]

For all but the most limited raids an airborne force required more than its personal weapons to make any significant impact on the enemy, whether in attack or defence. Indirect fire support was necessary to neutralise enemy positions prior to an assault and to disrupt and harass the enemy as they prepared to attack. Direct fire support, namely anti-tank weapons, was also required if an airborne force stood any chance of survival in the face of enemy equipped with armour. Anti-tank firepower was extremely limited during the early phases of development. Even during TORCH Frost complained that 1 Parachute Brigade was woefully equipped with only the Boys anti-tank rifle and the gammon bomb. These were clearly inadequate weapons with which to fight any sort of armoured force.[30] The only option under attack by armour was to withdraw and preserve the airborne force to fight again.[31] Issuing the Projector Infantry Anti-Tank (PIAT) gave paratroops some level of protection, but with the introduction of gliders, heavier anti-tank weapons could be delivered to the battlefield. Initially the 2-pounder anti-tank gun was allocated; however by 1942 this gun was not powerful enough to penetrate any but the lightest armour and it was never used operationally with airborne forces. The War Office quickly provided a heavier weapon and the 6-pounder anti-tank gun required only minor modifications to fit into the Horsa glider. A shortened axle width resulted in the 6-Pounder Carriage Mark Three.[32] The gun was issued to anti-tank sections within airlanding battalions and to specialist airborne anti-tank batteries of the Royal Artillery. It was first flown into action during HUSKY and used in every following airborne operation until the end of the war. With its low profile and armour-piercing capped (later armour-piercing sabot) ammunition it was claimed to be the most effective anti-tank gun in service at the time.[33] The later and larger 17-pounder anti-tank gun's introduction into service fortunately coincided with that of the Hamilcar, the only glider capable of carrying it. As with personal weapons, the increase in firepower had inevitable consequences for the size and weight of the load, which in turn resulted in operational restrictions, in that the Hamilcar required a larger LZ than the Horsa. This principle is clearly demonstrated by the provision of indirect fire artillery to airborne forces.

The gun initially allocated to airborne artillery units was the 3.7-inch mountain howitzer. This gun had been introduced into service in 1917; it weighed 773kg and fired a 9kg shell 5,490m.[34] It was quickly realised that a weapon with a greater weight of fire would be required and the obvious choice to supersede it was the 25-pounder.[35] Its ubiquity would simplify the logistics of ammunition resupply and it had a vastly superior range of 12,253m. However the gun was more than twice the weight of the mountain gun and its ammunition was almost 25 per cent heavier. The 25-pounder required extensive modification before it could be flown in a Hamilcar and the effort to achieve this was considered excessive.[36] VARSITY was the only operation in which the 25-pounder was flown forward, when two guns were taken to fire smoke shells to spot targets for ground attack aircraft.

Fortunately, an appropriate compromise became available in 1942 in the shape of the American 75mm pack howitzer. The gun had been designed in 1920 but had only been recently modified by placing it on the M8 (airborne) carriage. At 610kg it was even lighter than the mountain gun and a third of the weight of the 25-pounder. The range of 8,930m was considerably shorter than that of the 25-pounder but more than 50 per cent greater than that of the mountain gun. The ammunition also weighed less than that of the mountain gun but was considerably less lethal than that of the 25-pounder.[37] Despite this, with its high level of accuracy, the 75mm pack howitzer was considered superior in all respects to the mountain gun and was issued to all British airborne artillery units.[38] The gun was available for operations with airborne artillery units by D-Day but only one battery (211 Light Battery Royal Artillery) was flown in during Operation MALLARD on the evening of 6 June 1944, the remainder coming ashore with the 'sea-tail'.[39]

In order to ensure that air transport was utilised in the most efficient manner the conflict between size and weight and firepower was pragmatically resolved in favour of the former. This is an important point as this compromise was only possible because the War Office owned the requirement and the resources to meet it. Due to this fact the General Staff could accept a reduction in firepower in the knowledge that it would expedite the acquisition process.

In order to be effective, firepower had to be matched with mobility. It was widely accepted that airborne forces' advantage in strategic and operational mobility quickly disappeared once the troops landed.[40] Their tactical mobility was severely limited, particularly during the first half of the war. During TORCH in particular a lack of transport was critical in not allowing the parachute units to carry sufficient fire support or to evacuate casualties.[41] Efforts to increase the individual paratrooper's mobility were relatively simply achieved. Folding trolleys, bicycles and motorcycles were easily procured and simply delivered by container and parachute or glider. However, the fire support element required something more capable to move guns and ammunition. This also applied to headquarters and supporting arms such as engineers and medics that would also have heavy equipment to move once on the battlefield. In 1941 the vehicles selected to fulfil this role were the militarised Austin Seven and Austin Ten and the

15-cwt truck; however the Austins were underpowered for purpose and did not see operational service.[42] The Willys Jeep arrived, via the lend lease agreement, in time for minor adjustments to be made to the Horsa design to facilitate its carriage. It was in service with airborne units in time for HUSKY but only a few were deployed by glider during the operation, mainly to tow the 6-pounders.[43] The Jeep could also tow the 75mm guns and experiments were carried out to maximise its operating capacity. A shorter working life was accepted for the vehicle in order to allow it to carry and tow greater loads.[44]

Each arm and service modified the Jeep to fit its own purposes. Medics fitted racks to the front and rear on which stretchers could be placed to assist in the evacuation of casualties. One soldier recorded 'those Jeeps certainly took some punishment too, for they, like us were loaded to full capacity. We were kept busy for weeks making modifications to those Jeeps to adapt them for Airborne [forces].... Those Jeeps were the most useful and adaptable trucks produced in the War without a doubt.'[45] The Jeeps were equipped with trailers to increase capacity. For the artillery, carriage of ammunition was a priority and each 75mm gun was served by two Jeeps and three trailers carrying a total of 137 shells per gun.[46] The airborne REME divisional workshop managed to fit all of its required equipment into five trailers including generators, lathes, drills, a water still, welding equipment and more.[47]

Despite its success the Jeep was not capable of towing heavier equipment such as the 17-pounder. More powerful vehicles were required and the 30-cwt truck was used as a tractor. However, use of this vehicle, like the 17-pounder itself, was predicated on the arrival of the Hamilcar. The Hamilcar had been designed to carry two Bren Carriers as a secondary requirement.[48] These multi-purpose tracked carriers were issued down to battalion level where they did valuable work taking forward ammunition and bringing casualties to the rear.

The ultimate expression of mobility combined with firepower on the battlefield is the tank and as early as the autumn of 1940 the War Office had decided that, to be sufficiently effective, any airborne force would have to contain a proportion of armoured fighting vehicles.[49] By mid-1941 the Vickers Light Tank Mark VII had been identified as suitable for use by airborne forces and the heavy glider specification was to be designed around it.[50] The history and use of the Mark VII light tank, known as the Tetrarch, and its airborne successors has been recorded in detail and does not need to repeated here. However, the development of the airborne tank does demonstrate another important principle that the War Office adopted.[51] By the end of 1941 the Tetrarch, which had first been produced in 1938, was considered obsolete and it was accepted that development of a suitable replacement would be undertaken by the US. The T9 (or M22) Locust light tank was proffered as the successor and, being slightly smaller than the Tetrarch, would fit inside the existing Hamilcar design. Concurrently Vickers-Armstrong were developing the design of their Mark VIII light tank, which became known as the Harry Hopkins. This design was better armoured and had a marginally more powerful engine than the Locust but was significantly wider and heavier by one ton. The Harry Hopkins began production in June 1942 and, as the successor to the Tetrarch, was put forward as a suitable airborne tank.

By September 1942 Britain was still waiting for delivery of the first Locust. Gale wrote to Renwick suggesting that as the War Office had no evidence as to the suitability of the Locust, the Harry Hopkins should continue to be considered as a future airborne tank. Because its increased width and weight meant it could not be accommodated in the Hamilcar further investigations were required to look at alternatives to lift the tank.[52] The Air Staff considered it absurd to even contemplate designing a new glider exclusively to carry the Harry Hopkins and in any case it would take at least two years to produce a new model.[53] The Air Ministry saw no justification to proceed with the design of another glider and in the face of the evidence the War Office had to concur.[54] Gale could either accept the fact that the Locust might not meet all of the airborne forces requirements but receive it now, or push for the marginally better Harry Hopkins but recognise that it could not be lifted for another two years. It was nearly always more efficient to accept and adapt current equipment to fit existing aircraft envelopes than it was to develop bespoke equipment which might then necessitate major aircraft design work. Once again the alignment of requirement and resources together in the War Office meant that compromises could be accepted to enable efficient acquisition.

The airborne tanks never really achieved their predicted potential. The Airborne Armoured Reconnaissance Regiment with twenty Tetrarch tanks was landed in the Orne bridgehead on D-Day and a handful of Locust tanks were landed on the east bank of the Rhine during VARSITY but neither detachment had any significant tactical impact.

The final class of equipment to be examined presented distinct problems to airborne forces. The development of electronic equipment was still relatively immature at the outbreak of war. Radio communications would be vital to airborne troops who might find themselves scattered on a DZ and need to concentrate and coordinate their actions. The state of radio technology throughout the war made the provision of communications to airborne forces problematic and has since become the focus of attention in a good deal of the popular historiography. Radios of the period relied on valve and crystal technology and were inherently fragile.[55] Ensuring that the equipment arrived intact, having descended by parachute or glider, was not an easy proposition. Even with specialised containers it had to be expected that a large proportion of radio sets would be damaged during delivery. As experience accumulated through exercises and training this led to radios being dropped in quantities two or three times higher than that actually required to ensure enough working equipment to service a workable network. As with the other equipment already described, the provision of radios necessitated a compromise between capability, size and weight. Although this was not necessarily a major factor between the radio sets themselves it became an issue when power was taken into consideration. As an example the No.19 High Power (HP) radio set was thirty times more powerful than the No.22 set and consequently could communicate over five times the range. Both radios were of comparable size and weight and both were Jeep-mounted. However, the No.19 HP radio had a higher power consumption than the No.22 and a consequent requirement for more batteries and bigger, heavier charging equipment. Therefore, before MARKET GARDEN it was considered that these size and weight constraints outweighed the No.19 HP set's obvious tactical advantages and hence only two sets were taken on the operation as opposed to dozens of No.22 sets.[56]

Several memoirs illustrate failures in communications and their subsequent effect on operations.[57] The official history also stressed the fact that radio sets of the highest possible power should have been provided as one of the major lessons to be taken from airborne operations during the war.[58] However, airborne forces were hampered by exactly the constraints that impinged on the rest of the Army. As the Royal Signals official history clearly identifies:

> The ideal [radio] equipment [for airborne forces] had not only to give a higher performance than standard types, but had to be more portable or more easily transportable and at the same time be capable of withstanding very much rougher handling. Since these very same ideals had for many years been striven after to a large extent in the design of all new standard equipment there could be no rapid solution.[59]

Where problems were blamed on inadequate communications they were in fact little different to those being experienced by the rest of the Army. At Arnhem during MARKET GARDEN, where communications constraints were identified as a factor in the overall failure of the operation, the culpability rested more with procedural shortcomings and a failure to adequately mitigate tactical risk than it did with technical equipment limitations.[60]

Whereas radio communications were often seen as a constraint during operations, alternative electronic equipment was recognised as an indispensable enabler. From April 1940 Bomber and Coastal Commands had been developing a radar beacon and transponder to assist navigation, to allow aircraft to automatically locate a position on the land or sea and to facilitate a degree of identification for friendly aircraft. The corresponding pieces of equipment were code named Rebecca and Eureka. Towards the end of 1941 the AFEE began to see the potential of the system to aid the location of DZs and LZs by approaching aircraft particularly at night or in poor weather. The AFEE approached MAP requesting collaboration on development, and soon after, the first trials were successfully conducted.[61] It would appear from the primary evidence that from this point onwards the AFEE took the lead in further development ahead of Bomber and Coastal commands. The system was released for use on special operations in April 1942 and was first fitted to airborne forces aircraft in the August of the same year. The Eureka beacon fitted into the normal paratrooper's kit bag and was carried down onto the DZ by advance paratroopers (pathfinders) where it was then assembled and switched on in order to guide in the rest of airborne formation. An aircraft carrying the Rebecca transponder flying at 2,000 feet could expect to home in on the beacon from eight to twelve miles away.

The system went through a number of iterations and became an essential aid during airborne operations.[62] Where the equipment was not used, failed or broke, the results were often significant if not disastrous. The failure of FRESHMAN in November 1942 was directly attributed to the malfunction of Rebecca in the aircraft involved.[63] During HUSKY in July 1943 the system was not used as the equipment did not arrive in North Africa in time to be fitted to the aircraft. Although its absence is not directly referred to in the post operational report it can be assumed that it did have an effect. Only 39 of the 104 aircraft carrying 1 Parachute Brigade that reached the area of the DZ managed to drop their paratroopers within half a mile of the target.[64] Again, in Normandy, smashed and faulty Eureka beacons led to very scattered drops in the dark during the early hours of D-Day. 9 Parachute Battalion was dropped in an area of over fifty square miles and could gather less than a third of its strength for its attack on the Merville Battery.[65]

Despite these minor setbacks the provision of support equipment to British airborne forces was generally successful. Excluding the references to the lack of heavy firepower and mobility during TORCH there are few accounts of deficiency or inadequacy having any direct impact on training or operations. It has to be recognised that much of the equipment required was a variation of that already in service and therefore the technical and engineering challenges were relatively simple. Notwithstanding this the main factor behind the achievement was that the War Office owned the requirement and the majority of the resources to fulfil it. This allowed the General Staff to compromise, improvise and allocate effort and materiel as necessary in order to meet the aim. Where the War Office did not have effective oversight of the supply process, such as with parachutes and containers, it appears probable that the requirement was overestimated,

Equipment and Technology 65

either inadvertently or deliberately and therefore shortages and delays never caused difficulty.

Underpinning the equipment development procedures the General Staff adopted two important, linked principles. First, a less than ideal level of equipment capability, especially firepower, had to be accepted in order to meet the constraints of weight and size imposed by transport by air. Second, it was more expedient to adapt and modify current in-service equipment to fit existing aircraft than it was to produce bespoke airborne equipment that might then require major aircraft design work. With these principles established the General Staff investigated and accepted several different methods of procurement. As an early measure existing equipment was accepted into service with airborne forces even though it was clearly unfit for purpose or obsolete. The adoption of the 2-pounder anti-tank gun and the Austin Seven are examples of this process, which allowed airborne troops to develop training, tactics and procedures while more suitable equipment was procured or modified. Development was accepted from the bottom upwards and innovation introduced in front line units was often adopted as standard equipment. This widened the experimental base to encompass much of the pioneering work that was taking place during training and even on operations.

Where new or improved equipment was required the General Staff recognised the value of collaboration in order to expedite procurement. The airborne community worked with the other services, for example Coastal and Bomber Command RAF in the case of Rebecca and Eureka, and with other nations such as the US in the case of the 75mm pack howitzer and the Jeep, in order to bring suitable equipment into service with reduced effort. The link with American airborne forces was very important in this respect and was developed throughout the war. From 1943 regular liaison letters were exchanged between the British Army Staff in Washington and the War Office dealing specifically with the development of airborne equipment.[66] This collaboration was reciprocal and America took an increasing interest in British developments although the method of liaison was not flawless and significant advances on one side of the Atlantic were still occasionally overlooked on the other as late as mid-1944.[67] Despite these lapses the War Office had created a system of procurement for airborne equipment that had multiple points of supply and therefore where one failed others were still available to be drawn upon. Due to this approach, the relatively undemanding technical and engineering challenges and, most significantly, the ability to direct the allocation of resources, the provision and quality of equipment to support airborne forces on the ground did not adversely impinge on training or operations during the course of development.

The Availability of Support and Transport Aircraft

In contrast to the situation surrounding personal and support equipment, the War Office had very little control over the provision of aircraft to support airborne forces during training and operations. The General Staff were entirely reliant on the attitude and efforts of the Air Ministry and MAP. By their own admission, the Air Staff's attitude to the provision of transport aircraft during the early part of the war was 'based more upon expediency than in conformity with our desired principles.'[68] They regarded the Army's constant requests for air transport with irritation. Vice Chief of the Air Staff (VCAS) Air Marshal Sir Wilfred Freeman wrote to CAS stating, 'I am unwilling that the Army should raise the question of special transport aircraft, modification to bombers, and so forth. Special transport aircraft is "buzzing in the General Staff's bonnet" at the present time, and if we are not careful we will find a big proportion of our heavy bomber capacity set aside for the production of this particular "bee".'[69] Despite this attitude the Air Ministry was proactive in providing guidelines for support to airborne forces early in the development process. Two principles were laid down. The first determined that aircraft could not be provided solely for the development of airborne forces and that parachute dropping had to be an alternative role only. The second principle stated that aircraft used in training should also be available and capable of being used during airborne operations. It would be pointless to train with one aircraft, no matter how suitable, if it could not be employed during operations.[70] This pragmatic approach appears to contradict the Air Staff's active resistance to airborne development; however their motives were not entirely benevolent and having set these criteria they then maintained that there was only one aircraft that could fill them.

Six Armstrong Whitworth Whitleys had been sent to the CLS at Ringway shortly after its establishment in June 1940.[71] It was admitted that the Whitley was technically far from ideal. The paratroops had to jump through a hole in the floor, a difficult and sometimes dangerous procedure. The best method of exiting an aircraft was through large doors in the side of the fuselage, as preferred by the Germans with their Junkers 52. This was recognised and the Air Ministry examined side door options, including the de Havilland Frobisher and Flamingo (and its military variant the Hertfordshire) and the Bristol Bombay. None of these aircraft were deemed to be suitable due to obsolescence or technical difficulties concerning the size and position of the exit door.[72] Despite these aircraft types being discounted at this early stage their suitability continued to be a matter of debate. The Hertfordshire in particular continued to be offered as a suitable platform for airborne forces by those closely involved in development. With some modification it was believed that the aircraft would be satisfactory and that the requisite numbers could be found for small-scale operations; however, it was accepted that modified bombers would still be required for large-scale operations.[73] In this case, in order to meet the Air Ministry's second principle two very different types of aircraft would have to be operated in the training environment. This burden was not acceptable at this early stage of development. It has also been suggested that there was no good reason why the Handley Page Harrow could not have been utilised but the Air Ministry's figures show only fifteen of these aircraft available in Britain in early 1941.[74] Moreover, MAP's prognosis for aircraft production did not include the Hertfordshire or the Harrow. Therefore while there might have been sufficient aircraft for small-scale operations and limited training there would have been no possibility of sustaining or replacing damaged or destroyed aircraft. The case for the Hertfordshire and other similar aircraft types ebbed away.

What then were the alternatives? The ideal scenario would be the procurement of a bespoke aircraft specifically designed for airborne operations. A set of key specifications had been listed: '(i) Long range; without auxiliary tanks. (ii) Low stalling speed. (iii) Accommodation for at least 10 men with their equipment. (iv) Easy exits (side doors being the best). (v) Availability in large numbers. (vi) Should be armed for its own defence. (vii) Should have bombcells to carry containers.'[75] However, bearing in mind Beaverbrook's production priorities the probability of getting such a requirement acknowledged and then having the aircraft designed and manufactured was low. The Air Ministry maintained that for some considerable time a specialist aircraft suitable for parachute operations could not have been developed and produced in quantity without serious prejudice to the production of aircraft required for other purposes.[76] MAP, supporting the Air Staff in this case, was entirely opposed to designing a new transport aircraft and preferred old and obsolete bombers to be used for transport and thus for airborne forces. Despite the associated technical and engineering challenges being relatively low, the ministry cautiously estimated four years as the time taken to produce a completely new type of transport aircraft.[77] By 1942 British manufacture of transport aircraft was negligible and arrangements for the production of future transport aircraft had not even reached the design stage. It was judged unsound to initiate plans for the production of transport aircraft in Great Britain at this point in time.[78] It was a question of priority, and since the Air Ministry and MAP controlled the resources the War Office could do little to change the situation.

Since the prospects of indigenous production of a bespoke aircraft were negligible in the short and medium term another option was required. Portal believed that the most promising method of achieving their needs would be to exert pressure on the Americans to expand the production of their transport aircraft, rather than to attempt to modify the British production programme.[79] Keyes had already identified the American Douglas DC-3 as a suitable aircraft for airborne forces early on in the development process.[80] The military variant of this aircraft, the Douglas C-47 (or Dakota to the British), was part of an American programme that the Air Ministry estimated was planned to manufacture 7,000 transport aircraft in the short to medium term.[81] The COS Committee agreed that the Dakota was probably the most suitable aircraft available, but towards the end of 1942 Portal was forced to admit that it would probably be some time before the RAF received sufficient of them – even if the Americans could be persuaded to allocate them. He proposed to wait and deal with the requirement when the British put forward to the Americans their entire needs for aircraft for 1943 onwards. It was clear that even by mid-1943 there would be insufficient Dakotas, or equivalent aircraft, with which to drop and tow the projected numbers of airborne troops and gliders required for future operations and training. Therefore prior to that period, as Portal had to admit, 'there is, I'm afraid, no alternative to the Whitley.'[82]

In mid-1940 the Whitley had been singled out as one of five aircraft types on which the main effort of manufacture might be concentrated.[83] Armstrong Whitworth was, with other manufacturers, instructed to continue production on current aircraft types, such as the Whitley, rather than increasing the manufacture of new types like the Manchester.[84] Despite this order the Air Ministry and others regarded the Whitley as already obsolete in 1940.[85] The RAF was eagerly anticipating the arrival of their modern four-engine bombers. Agreeing to release Whitleys was therefore a means of meeting the requirements of the Prime Minister's order without seriously jeopardising the Air Staff's immediate plans for expansion of the bomber force. Very early on in the development process the Whitley was identified by those having to operate with it as 'thoroughly unsatisfactory'. Keyes pointed out its failings to Churchill.

> [The Whitley] can carry only 8 men, who would have to sit throughout the passage overseas, huddled up in the bomb tube in great discomfort, and then drop through the middle of a small hole, with no margin for error in poise, conditions which are calculated to damp the light hearted enthusiasm with which these young men volunteered for a hazardous adventure.[86]

The first fatality after only a few weeks of training at Ringway did nothing to improve that enthusiasm. The first few Whitleys used for airborne training had been retrospectively fitted for the task in an *ad hoc* manner. From the spring of 1941 the Whitley V came into production with integral fittings for airborne operations and the capacity rose from eight to ten men. Notwithstanding this improvement, it was projected that if every available Whitley was committed to airborne operations, only 600 to 700 paratroops could be dropped, a figure far short of Churchill's 5,000.[87] If airborne development was to maintain a desired rate of progress more aircraft would be needed. With no likelihood of receiving specialist transport aircraft before 1943 the only viable option was to increase reliance on the bomber force.

Naturally the Air Ministry resisted the suggestion that more bombers might be marginalised by their use by airborne forces. It maintained that it had already investigated the use of the Shorts Stirling, Handley Page Halifax and the Manchester for airborne operations and had found them incompatible. It reluctantly conceded that the CLE could conduct its own investigations into the suitability of these new types of bombers as long as this did not involve any permanent allotment of aircraft to Ringway and the work did not take precedence over operational requirements.[88] The Air Staff was acutely aware of the interference that continuous modifications caused to the flow of aircraft production.[89] The Air Ministry might have been expected to resist this measure more strongly, although it probably knew that passive resistance would be enough to prevent the allocation of bombers to airborne forces seriously impinging on the strategic offensive.

Converting or modifying any aircraft once it is in service, while not technically difficult, presented serious engineering challenges. As the Air Ministry's Director of Operational Requirements (DOR) observed, it was a 'problem which involves considerably more than providing seats for the men and space to accommodate equipment.'[90] Apparently minor alterations took up a disproportionate amount of time and effort and many were required before a bomber could be used to drop paratroops. Lights in the cabin to indicate the time to drop had to be fitted and connected to the cockpit, intercom from cabin to cockpit was required, rigging lines had to be fitted for parachute static line attachment, floors had to be raised and reinforced and a dispatch hole had to be built. The process could be accelerated by purchasing 'off the shelf' equipment that could be temporarily fitted and removed as required. However, in most areas progress was tortuously slow; for example it took eight months to make a decision on the type of rubber matting to fit in the cabin in order to make the flight more comfortable for the paratroops.[91] Additionally there was one overriding consideration that could render any modifications futile no matter how carefully considered. DOR commented once again.

> It should be impressed upon the War Office that aeroplanes are delicately balanced craft which cannot be loaded like Army lorries with space as the only limitation. Carefully prepared loading with scientifically designed stowages to limit movement during flight are indispensable. It requires much elaborate calculation trials and technical investigation and the preparation of numerous varied loading tables for each type.[92]

Conducting centre of gravity trials required a huge amount of technical staff effort and a lengthy period of time, particularly when the Air Ministry was reluctant to release aircraft.[93]

The CLE was first given clearance to begin trials on the Vickers Wellington Ia, Ic and II in May 1941. This work was to cover 'technical details of the structural alterations necessary to convert the Wellington from normal operational work to the role of paratroop dropping' and 'air tests with dummy and live troops and full load tests to determine the effect under all conditions of the centre of gravity.'[94] It was not until June 1942, thirteen months later, that the trials were complete and the Wellington was finally given clearance to conduct parachute operations and training.[95] Even then the aircraft was only cleared to carry and drop eight paratroops, a capacity inferior to that of the Whitley at that time. In response to this 1 Parachute Brigade ran its own in-service trials a month later and concluded that it was perfectly practicable to operate sticks of ten men with four containers out of the Wellington.[96] Commander 1 Parachute Brigade urged the AFEE to verify the findings and amend the official aircraft release, which was eventually issued in November 1942.[97] Therefore the entire process took eighteen months.

Even when an aircraft was finally cleared for parachuting it did not necessarily result in improved availability for training and operations by airborne forces. The parachute role was always considered secondary; even the 'obsolete' Whitley had been primarily allocated to Coastal Command for anti-submarine operations. The Air Ministry maintained that it would be difficult to concentrate parachute-dropping Whitleys in one or two bomber squadrons without involving a double move for each replacement aircraft. It was considered that the administrative difficulties of concentrating parachute aircraft in one or two squadrons outweighed the training and operational advantages.[98] As Whitley Vs left the production line they were allocated wherever the Air Ministry believed the priority lay. This meant that the modified aircraft were spread across the RAF and were difficult to concentrate for training or operations. The situation was only alleviated when dedicated RAF formations were committed to airborne operations, although even then the aircraft were frequently committed to other activities.[99]

The technical and engineering challenges of the modification programme were exacerbated by organisational limitations. Initially work was restricted by the improvised nature of the technical trials, development and experimentation establishment at the CLS. When the CLS was expanded to the CLE an Experimental Flight was established alongside the TDS.[100] These departments, responsible for all trials, were still based on a minimum and clearly inadequate level of manpower. The TDS, run by a civilian Chief Technical Officer, was split between two productive sections, one run by a Senior Technical Officer (STO) and the other by an Engineering Officer. Under the former was a small trials section, just a handful of civilian and military officers to prepare, set up and report on aircraft trials. The Engineering Officer had one subordinate officer, Flight Lieutenant Pithkelly, to work as Contractors' Liaison and Modifications Officer, a single man to act as the point of contact between the CLE and the numerous manufacturers nationwide who would have to embody all the modifications that had been trialled and recommended. All of those trials had to be carried out by the Experimental Flight made up of a Flight Lieutenant and a handful of pilots.[101] This small organisation struggled to cope with the mass of experimental work that was required in these early stages of development.

With the expanded establishment of the AFEE the Engineering Officer gained his own workshop with 118 technicians. There was also a separate dedicated Design Section, and Glider Trials also had their own TDS. Flight Lieutenant Pithkelly was now one of two officers responsible for parachute aircraft technical instructions. The same section had dedicated officers working on aircraft loading and accommodation and aircraft calculations. This section now had its own experimental flight to fly trials solely to test parachute aircraft modifications.[102] The trials process does appear to have improved in the second half of 1942 as a result of the formation of the AFEE. The level of activity certainly increased as the Albemarle, Lancaster, Halifax, Manchester and others were all cleared for airborne operations between mid-1942 and spring 1943.

In addition to internal organisational limitations, the CLS, CLE and AFEE also had to deal with inter-departmental friction. This was essentially due to a lack of coordination across the many ministries and departments necessarily involved in developing aircraft capable of being used by airborne forces. With the Air Ministry, the War Office, MAP and the Ministry of Supply all involved no single ministry was given a clear lead role.[103] This led to nugatory work being carried out by both MAP and the War Office as each independently tried to coordinate the aircraft modification programme. MAP was directing work to be carried out by aircraft manufacturers without any clear knowledge of the War Office's detailed requirements for airborne forces. Concurrently, the War Office was making assumptions in its planning of airborne development without any technical experience of aircraft capabilities. The CLE was the natural point at which these conflicts could be resolved, but prior to the formation and separation of the AFEE the establishment at Ringway was inadequate to coordinate the often opposing standpoints.

Given the nature of the dilemma it is not surprising that the Air Ministry and MAP gained the upper hand over the War Office, with the Airborne Division being portrayed as offenders. The MAP technical staff complained of a general lack of coordination and were doubtful that production would be ready in time. There was also a feeling that the War Office did not fully appreciate the limitations in loading placed on the whole modification project because of the essential need to balance the aircraft. MAP was aware that a flood of fresh requirements would likely ensue as the Army began to obtain more experience in the field.[104] The Airborne Division was accused of trying to bypass the official process for conducting trials and getting modifications accepted.

> Both the Ministry of Aircraft Production and the Air Ministry are frequently faced with most embarrassing situations owing to the enormous amount of 'backdoor' business which goes on between the various Ministries and between the Army and Air Force Departments and Headquarters concerned with Airborne Forces and their equipment. As often as not these backdoor approaches are made with the object of getting something done more quickly than could be expected by the official channels and equally frequently the essential partner is short-circuited. The result of course is confusion. With three Ministries concerned, any departure from official methods requires very careful handling. There also seems to be a strong belief among those concerned with the operation of Airborne Forces that ordinary Service technical procedures are not applicable to their novel equipment.[105]

Considering 1 Airborne Brigade's independent trials with the Wellington this criticism appears justified. The AFEE was warned against accepting these approaches and was ordered to cooperate with the Airborne Division in an advisory role only. They were able to discuss projects with the Division and advise them on the practicability of their ideas. However, the AFEE was not empowered to take on experimental work on behalf of the Airborne Division.[106] An attempt at the end of 1941 to formalise the methods and responsibilities for trials and conversions had not yielded any benefit.[107] Once again the formation of the AFEE appears to have improved the situation and complaints and comments receded from mid-1942. At the end of 1942 DMC accepted his department as being formally responsible for coordinating all technical airborne requirements, stating that only those requirements placed with DMC would be accepted and placed on the official list. DMC sought to reassure the War Office whilst warning them against continuing to bypass the system when he wrote 'I realise that a good deal of useful development is obtained by direct contact between members of the Airborne Division and manufacturers. While not proposing to damp anybody's enthusiasm, I suggest that no official cognisance should be taken of requirements which do not come through official channels.'[108] The War Office accepted this approach and adhered to DMC's suggested method of categories and priorities for organising airborne technical requirements.[109]

Regardless of these improvements in organisation and coordination it was becoming clear that reliance on the bomber force to transport and drop Britain's airborne forces was going to prove highly inefficient and was not going to achieve the numbers required. The COS Committee was appraised of the situation in April 1942. By this time the Whitley was considered obsolete and only suitable for training, hence it was omitted from any projected calculations. The Wellington and Lancaster conversion programmes therefore determined the numbers of paratroops that could be dropped during an operation. It was estimated that by 1 May 1942 enough Wellingtons would be converted to drop 500 men, this number rising to 820 by 15 May. By 1 June 1,150 paratroops could be dropped by Wellington with a further 450 being dropped by Lancaster, although this was subject to successful completion of trials. Finally, it was projected that by 1 July a total force of 2,700 could be lifted and dropped, 2,000 in Wellingtons and 700 in Lancasters (still subject to the trial results).[110] Although this figure met the War Office requirement of being able to operate 2,500 paratroops by 31 July 1942 it still only equated to a single parachute brigade and required twenty squadrons of heavy and medium bombers to achieve it.[111] As Britain's airborne forces continued to grow towards a force of more than two divisions a more efficient means of transport had to be found.

Douglas transport aircraft had already been identified as ideal airborne aircraft and the US was acknowledged as the most effective source for procurement. An initial proposal for procuring Douglas DC-2s and DC-3s involved Churchill trying to influence the Dutch government to put pressure on KLM to donate their few Douglases stranded in Britain. It was also estimated that at least 1,500 Douglas aircraft were employed in civil aviation worldwide, and as an alternative to the KLM plan consideration was given to purchasing a number of them.[112] Despite Churchill agreeing to approach the Dutch government these initial attempts to obtain Douglas aircraft were unsuccessful. Even if a number had been secured the Air Staff believed that their employment on the Trans-Africa route was of greater military importance than support to airborne forces.[113] In fact the first American aircraft to be used by British airborne forces were not from Douglas. The Consolidated Liberator and Lockheed Hudson were the first candidates for airborne duty in mid-1941.[114] By October of that year examples of each aircraft had been secured for trials.[115] However both aircraft suffered from the same limitations as the British bombers. Neither was designed as a transport aircraft and both required extensive modification, which yielded only a poor capacity of eight to ten paratroops.[116]

It was the Douglas aircraft's capacity that made it particularly suitable and attractive for airborne operations and with the production of the C-47 Dakota a military variant was available. All that was required was to persuade the US to hand them over to the RAF in sufficient quantities. Initial enquiries into the possibilities of aircraft production in the US began very soon after Churchill became Prime Minister but in mid-1940 these were mostly informal and did not include any reference to transport aircraft, concentrating instead on fighters and bombers.[117] On 7 December 1940 Churchill wrote to Roosevelt stating that Britain could no longer pay cash for war supplies and asked the President to extend aid beyond that which could be immediately paid for. Congress passed the Lend Lease Bill in January 1941.[118] The Lend Lease programme would not bear fruit for many months and in the interim negotiations were taking place to release to Britain aircraft that had been financed under US Army appropriations. Slessor chaired these negotiations for the British, beginning in December 1940. The 'Slessor Agreement' apportioned over 25,000 aircraft, worth nearly 2 billion dollars, to be delivered by the Americans to Britain by June 1942. Only 147 transport aircraft, all Dakotas, were included in this total.[119] In fact these figures were 'unduly optimistic' and monthly production in the US did not achieve even half that required to reach the projected totals by June 1942.[120] With transport aircraft afforded such a low priority their production was bound to suffer. In fact, there is no evidence that any of this initial order for Dakotas was delivered.

Further bids for transport aircraft do not appear to have been made until the spring of 1942. In April, CAS stated that 312 transport aircraft had been ordered from the US for delivery during 1942. A month later Portal assured Browning that 650 aircraft would be delivered during the year but contradicted himself by warning those involved in airborne development that they would be deluding themselves if they imagined that these aircraft would be available by the end of 1942.[121] The first Dakota equipped for parachute dropping would not be delivered to the AFEE for initial trials until August 1942 at the earliest and as the initial trickle of aircraft began to arrive, those trials imposed further delays.[122]

To date British paratroopers had been despatched through a hole in the floor of their aircraft. The side door exit of the Dakota required minor modifications and then extensive trials in order to ensure that British parachute equipment, which differed from American, could be operated safely. New parachute drills had to be written. This work took place between August and October 1942, although the final report was not released until March 1943.[123] The consequence of these delays was that the Dakota was not available for operations with the RAF in appreciable numbers until the formation of No. 46 Group on 17 January 1944. This fact, coupled with a lack of suitable bombers available in North Africa at the end of 1942, meant that during the airborne operations as part of TORCH British paratroops had to jump from United States Army Air Force (USAAF) aircraft of 51 Wing. These operations were characterised by poor air planning, confusion on the mounting airfields and drops that were scattered and not entirely accurate. Despite their best efforts much of this can be blamed on 51 Wing, which did not have sufficient experience with British troops and had an inadequate combined airborne-air ground base organisation.[124] It was largely luck that prevented higher casualties on the drop zones.

Equipment and Technology 75

The flow of aircraft to Britain steadily increased to monthly deliveries of twenty-five to forty-five aircraft from the end of 1942 and into 1943.[125] However the RAF Dakota never quite achieved the ubiquity with which it has often been credited.[126] Taking the 1 British Airborne Division fly-in to Arnhem during MARKET GARDEN as an example, on initial examination it would appear that the Dakota bore the bulk of the workload with approximately 60 per cent of the 1,089 aircraft that flew troops and equipment in being of that type. Among the aircraft dropping paratroops dominance was almost total with 97 per cent of the aircraft being Dakotas. However all of these aircraft came from the USAAF; none were flown by the RAF. When the figures for re-supply flights are scrutinised the picture is different. Of the 611 resupply sorties flown by the RAF less than 40 per cent were flown by Dakota, the rest being by Stirling.[127] The situation was similar with glider towing sorties, with only 36 per cent being flown by Dakota, the remainder being flown by a variety of bomber types.[128] This arrangement of aircraft, with practically all paratroops being dropped by Dakota while bombers towed the majority of gliders, represents an efficient solution. In terms of carrying paratroops the Dakota was two-and-a-half times more efficient than the bombers in this role, while one aircraft, whether Dakota or bomber, was required to tow one glider. In fact this efficiency occurred more through accident than design. It had been decided that all paratroops across the entire operation would be dropped by the American Troop Carrier Command in order to simplify administration and the allotment of troops to airfields in England.[129] Troop Carrier Command only operated the Dakota. Additionally some gliders were too heavy to be towed by the Dakota and therefore they had to be paired with the heavy bombers.

If the provision of aircraft to support paratroopers was difficult then the challenge of providing aircraft to tow gliders was more so. In general terms the problems were the same: there was no chance of a specialist tug aircraft being procured, any assistance from America would not become apparent until the end of 1942 and therefore prior to that time RAF aircraft would have to be modified for purpose. However, two factors particular to supporting glider operations aggravated those problems further. First, the conversion of an aircraft for towing gliders was a more daunting engineering prospect than modification for parachute operations. The stress placed on an airframe by pulling a fully loaded glider weighing up to sixteen tonnes was phenomenal, and considerable structural modification was required, followed by lengthy trials. By mid-1941 design work was underway to devise a means of avoiding the removal of the rear turret in the Wellington III when it was being used as a glider tug. This was expected to take five or six weeks and a trial installation would then have to be carried out before the towing modifications could be incorporated in production aircraft.[130] Halifax and Lancaster trial installations were also progressing but it was estimated that it would be five or six months before production aircraft could be fitted for towing.[131] Second, the number of aircraft required for glider towing within an airborne formation was greater than that for parachute dropping. It was estimated that an airlanding brigade would require 25 per cent more aircraft than a parachute brigade.[132] In fact the Arnhem example suggests that the figure was more like 75 per cent.[133] Attempts were made to reduce this number by trialling double tows, i.e. two gliders behind one tug, but the experiments were not entirely successful, particularly for the heavy gliders, and the decision was made to adhere to one glider per tug.[134]

There was one prevailing issue that aggravated all other factors when dealing with the development of the airlanding component. Glider development lagged behind that of the parachute element. Essentially the Air Ministry, War Office and MAP were attempting to develop a capability that was a technological unknown. Due to a lack of practical knowledge and experience, mistakes were made which often did not come to light until they had become severe problems. As with parachute training and operations, the Air Ministry had allocated only the Whitley to support airlanding development. While there were advantages in operating only one aircraft type to tug gliders and drop paratroops, as heavier gliders were developed it was discovered that the Whitley was inadequate as a tug. Ismay considered that the provision of a suitable towing aircraft was not being pressed forward with the urgency it deserved and was concerned enough to urge Churchill to become personally involved.

The Prime Minister was deeply unimpressed to discover that the Whitley, the only aircraft with which the Airborne Division was equipped, had proved to be unsuitable for towing gliders, leaving Britain's airlanding troops with no aircraft usable for this purpose. He demanded an explanation.[135] CIGS outlined the General Staff's view that the Whitley aircraft could not tow a fully laden Horsa owing to a defect that meant the engines overheated when in this configuration. This resulted in training being confined to Hotspur gliders. This was not satisfactory due to the Hotspur being designed for glider pilot training only.[136] CAS attempted to paint a less negative picture of a situation, which, he justifiably claimed, was a result of the very little practical experience of glider towing. He did not however consider that the Whitley was entirely unsuitable for training purposes. He conceded that it could not tow a fully laden Horsa without overheating but was content that MAP was trying to improve the engine cooling and that their efforts would be helped by the colder weather of winter. Churchill was unimpressed by Portal's explanation and noted: 'There may well be a scandal about this. What action do you propose?'[137] The Minister of Production, Oliver Lyttelton, supported the Air Staff in this case stating that it was quite untrue that the Whitley could not take off with a fully loaded Horsa. MAP had to admit that the operational range

Equipment and Technology

would be reduced to 180 miles but was confident that this was sufficient for training purposes.[138] However, the Whitley was patently not sufficient for use during airlanding operations, restricted as it was by climate and range, and therefore, following the Air Ministry's own criteria, it was also unsuitable for training.

One thing all the ministries involved did agree on was that the answer to the glider-towing problem lay with the Dakota. However, the wait for initial deliveries of the aircraft had the same impact as it had on parachute development, and in a similar response alternative aircraft were pressed into service. The full inventory of heavy bombers was considered for conversion and many were modified, with aircraft becoming available from the end of 1941 onwards. In addition to these, more novel combinations were considered and trialled, including the Bristol Beaufighter towing the Horsa and the Supermarine Spitfire towing the Hotspur, although neither was used operationally.[139] By mid-1942 the specifications of standard British towing equipment had been forwarded to Douglas so that C-47s coming off their production line and destined for the RAF would have the necessary modifications built in.[140] The Dakota still required trials when it arrived in Britain, as the Horsa glider was considerably heavier than the standard American glider. By the end of 1942 the Dakota had only been cleared as a tug for the Horsa in 'British winter conditions'.[141] The numbers slowly rose and trials continued into 1943 but the Dakota was not operationally available in significant numbers until the beginning of 1944.

With Churchill insisting that HUSKY had to take place in June 1943 the British airborne element suffered in the same manner as it did for TORCH. Sufficient converted bombers could not be released either from the Middle East or from Britain and so the majority of 1 Airlanding Brigade had to be towed into battle by the American Troop Carrier Command. Of the 145 tugs that took part 75 per cent were American C-47s, with the remainder being made up of RAF Halifaxes and Albermarles. Only 34 per cent of the American-towed gliders successfully landed on Sicily, with over 50 per cent ditching in the Mediterranean.[142] Churchill believed that this constituted 'a serious disaster'.[143] There were several reasons for this poor performance, the chief among them being the unfamiliarity of the American pilots with airborne operations and a truncated training period, a situation forced by the lack of RAF aircraft and crews available in the theatre. The Deputy Supreme Commander of HUSKY, General Alexander, was in no doubt of the lessons that had to be learned. RAF pilots, crews and machines had to be made available and put aside for airborne forces with whom they should live, work and train to the exclusion of everything else.[144]

One result of the British airborne experience during HUSKY was that a lack of confidence in the C-47 Dakota began to manifest itself.

The C-47 is a very poor plane for transporting paratroops. It is helpless in the presence of armed enemy air of any description, one bullet in the gas tank will bring it down in flames and it mounts no weapons of any description, which has a bad effect on the morale of both passengers and crew. This nervousness was evidenced in many cases in HUSKY by the pilots' failure to reduce speed over the dropping ground with resultant increase in the hazard to the paratrooper when his parachute opened.[145]

The problems with dropping paratroopers also applied to glider operations over Sicily. Many gliders were dropped short in the sea because of tug pilots' reluctance to approach too close to the coast where the anti-aircraft fire was believed to be concentrated.

Another event also had significant impact on the use of the Dakota as a glider tug. The sixteen-tonne Hamilcar glider began to enter service in the second half of 1943. As training with the giant glider began in November 1943 it became clear that the Dakota did not have the power to tow it. Therefore at a time when, with the growth in numbers of the Dakota, parachute fittings could be discontinued in some bombers, glider-towing fittings had to be retained.[146] This decision was essential in order to enable the large-scale airlanding operations in North-West Europe in 1944 and 1945. The 36 per cent of Dakotas used to tow British gliders during MARKET GARDEN was the highest achieved, the figures for the British OVERLORD operations and VARSITY being 28 and 24 per cent respectively.[147] The majority of the workload was still borne by the modified bomber fleet.

The provision of transport aircraft during the Second World War was based on expediency rather than planned principles and this included support to Britain's airborne forces. The parlous state of Britain's aircraft industry immediately prior to the outbreak of war has already been described. Even when indigenous aircraft production was at peak capacity it never achieved its full potential. The industry suffered from the defects inherited from its small-time origins, weaknesses resulting from the haste of its expansion, the retarding force exerted by the trade unions and restrictive practices in manning and demarcation.[148] The aircraft industry appeared to have an inherent limit to its capacity. At peak effort human resources were strained and 'unpredictable and intangible influences' appeared to hold back output.[149] In the case of transport aircraft this was exacerbated by the very low priority that they attracted. The Air Ministry were utterly dismissive of the notion that a specialist aircraft type could be provided from national resources. By the time that the Air Staff realised the need to divert productive capacity from heavy bombers to transport aircraft it was too late.[150] The estimate of four years for production precluded any new type entering service in time to be of use. Instead the Air Ministry was relying on the production capacity of the US to fulfil the RAF's transport aircraft requirements.

However, the Air Ministry at the very highest level knew that this was an unrealistic hope in the short and medium term. Portal realised that very large increases in the output of the American aircraft industry could not be made in 1942.[151] Joint estimates of the quantity of aircraft that the US could supply to Britain were dangerously over-optimistic. When the American production programme failed to deliver transport aircraft at the agreed rate it caused a clash between the Air Ministry and MAP as to which was to blame. The Air Ministry prepared defensive briefs, concerned that an attempt might be made to pursue the matter back beyond the spring of 1942. The Air Staff could be accused of neglecting air transport during the early stages of the war and leaving MAP with no choice in 1942 but to accept production in the US. It required detailed examination to refute this thoroughly as the Air Staff had already accepted in 1941 that the main source of supply of transport aircraft would have to be the US. They were then left exposed when American production only allocated a very small number of transport aircraft to Britain during 1941. The situation never recovered as American production of transport aircraft lagged continually behind the projected programme.[152]

The general neglect of transport aircraft, deliberate or otherwise, suited the aspirations of both the Air Ministry and MAP and allowed them to proceed with their core plans for expansion in bombers or fighters respectively without being distracted. Anyone with a requirement for large scale transport by air, and this of course included Britain's developing airborne forces, was forced to adopt a 'make do and mend' strategy. The War Office could do little to influence the situation because, although it owned the requirement, the Air Staff controlled the resources to fulfil it. The Air Ministry's 'expediency' translated into a requirement for improvisation as less-than-suitable aircraft had to be pressed into service. Paratroopers and glider pilots were forced to adapt to different British and American procedures across multiple aircraft types.

The Development and Production of Gliders
If the provision of support aircraft to airborne forces proved to be challenging at least a solution was available. Aside from priorities and resource control, the principal difficulty in producing transport aircraft and converting bombers for airborne support lay with the engineering rather than the technical aspects. The desired result was known; it was the method of achieving it that proved somewhat tortuous. The staff grappling with the development of the glider were not so fortunate. They had to turn a nebulous concept with no precedent into solid reality within a short time. It was a massive technical challenge that also presented its own unique engineering problems. Before the war the Air Staff was not even certain whether gliders represented merely a sporting activity or might prove to be a novel weapon system.[153] The concept and doctrine of airlanding

operations utilising gliders was subject to continuous development throughout the war, the progress of which will be examined later. Suffice to say, at this stage, those responsible for the production of gliders had to do so against an ambiguous requirement that remained fluid for much of the war. The tactical and technical aspects of glider development were indivisibly linked; however the technical aspects will be dealt with first in this chapter.

The Air Ministry had shown some interest in gliding at least as early as 1922 when it sent Squadron Leader Wright to Germany to observe sport gliding trials. Wright's report concluded that no direct commercial or military value could be attached to the use of gliders.[154] This remained the Air Ministry position until the German assault on the Belgian fort at Eben Emael proved otherwise. As Britain embarked on its own airborne project early estimates indicated that an airborne brigade group would require approximately 1,340 tons of equipment to be lifted by air.[155] There were essentially three methods of delivering this equipment to the battlefield: by transport aircraft, dropped by parachute or in gliders. The Germans employed transport aircraft to deliver men and equipment directly in the assault in Norway, The Netherlands and Crete. However, this technique was incredibly wasteful and in the case of Crete in particular it resulted in high losses. Unless an aerodrome was seized and held, aircraft were forced to land on unsuitable ground, such as the beach at Maleme, normally resulting in the loss of the aircraft.[156] These aircraft losses were unacceptable even to the Germans with their surfeit of Junkers JU-52; to the British, with the dearth of transport aircraft already described, the tactic was inconceivable.

Dropping heavy equipment by parachute also brought its own challenges, both technical and tactical. The technical difficulties initially centred on the load constraints imposed by the available parachute technology. In mid-1941 the heaviest loads being trialled for dropping were crates of 25-pounder ammunition, weighing just over 50kg. Heavy air drop was another example of an essentially Army problem, supply by air to isolated units on the ground, that could only be solved by the RAF.[157] Because of this, progress in this area was slow, and it was 1944 before the means had been developed to drop the Jeep and the 6-pounder anti-tank gun from the bomb bays of the Halifax and the Lancaster. However the technique required an inordinate amount of effort to rig the loads to withstand impact on landing, and even then engineering problems still existed, such as parachutes that conflicted with aircraft antennae as they deployed.[158] Added to the technical challenges there were also tactical concerns. Both the 75mm gun and the 25-pounder could be disassembled to be dropped in several loads by parachute. The 75mm gun broke down into nine separate parachute loads and the 25-pounder into twelve. The twelve 25-pounder loads could be dropped from a single Dakota with ninety-six rounds of ammunition dropped from a following aircraft. The problem was the amount of time it then took to bring the gun into action as the loads had to be found, brought together and then the gun assembled. If one load was lost or damaged the entire gun was rendered useless.[159] The technical and tactical challenges combined meant that heavy air drop never became a common technique for British airborne forces, although it was utilised for limited special forces operations.

The only practical method of delivering large quantities of heavy equipment by air to the battlefield was therefore by glider, and early attempts to define the precise requirement appeared promising. Even before Churchill's minute in June 1940 it had become apparent that the prospect of towing large gliders behind bombers warranted further investigation.[160] An initial report by the Air Ministry's Director of Scientific Research (DSR) concluded that a glider with a useful payload of around one tonne was not an unreasonable proposition. These gliders, it was supposed, might hold four or five troops and could be towed normally in trains of three behind a suitable towing unit, although new aircraft, such as the Wellington, might be capable of towing up to five. Furthermore DSR reported that the time required to design and construct these gliders would not be excessive, indeed experimental designs were already being produced in mid-1940.[161] At a subsequent meeting DSR's technical prognosis was enhanced and amended by the application of military judgement. War Office representatives agreed that gliders with a capacity of seven to eight men would be tactically acceptable as this represented the size of a standard infantry section at the time.[162] Notwithstanding this concurrence senior Air Staff were adamant that no large production orders should be placed until prototypes had been cleared and the requirement clearly understood. The technical challenges should be addressed before manufacturing began, or as ACAS (Technical) succinctly put it, 'We must walk before we can run.'[163]

By the beginning of 1941 the War Office was beginning to complain about a lack of progress in airlanding development and the Air Staff had to concede that this was due mainly to a deficiency of training gliders.[164] Until this time the only training aircraft available were converted civilian, single-seat Kirby Kite sailplane gliders. In March 1941 it was confirmed that an order had been placed with the General Aircraft Company for 400 eight-man Hotspur gliders. It was made clear that these gliders were intended for training only.[165] This was fortuitous as concurrently it was decided that there were tactical objections to eight-man gliders. These included the obvious advantages of landing men compactly in larger numbers than eight, the difficulty in towing gliders in tandem through bad weather or at night, and the high number of towing aircraft that would be required to land large groups of men simultaneously. Added to this might have been the fact that smaller gliders would require a higher number of glider pilots to fly the same amount of men and equipment. Despite this it was decided that the Hotspur could be employed if an airborne operation was required before larger gliders were available.[166] Although the General Staff had explicitly stated that an eight-man capacity was inadequate except *in extremis*, the accepted training role of the Hotspur began to drift towards an operational purpose. By September 1941 a deficiency of larger gliders was still apparent and it appeared that the Hotspur would indeed have to be pressed into operational service.

Converting a training glider to operational use was not a simple task and problems associated with operating the Hotspur at a higher weight with tactical loads had already been identified. These included a low maximum towing speed, a high landing speed requiring a long run on, a cramped and fragile hull, heavy controls and problems with jettisoning the undercarriage. In addition to this it was doubtful whether the Hotspur could manage a full load of eight troops with equipment and it was estimated that a more normal load was likely to be five troops.[167] Trials with operationally loaded Hotspurs were conducted through September and October 1941, with a report being published in November. It was considered that the hull fragility and high landing speeds could be coped with but the fact remained that the capacity of five equipped soldiers was simply uneconomical. In the event the Hotspur was never used operationally although considerable staff and technical effort were expended in order to cover the contingency. The Hotspur remained the principal training glider and once introduced it effectively met the requirement for the initial training of glider pilots throughout the war. By the time gliders were required operationally (the first occasion being FRESHMAN in November 1942) it had been superseded by a larger, more useful aircraft.

The War Office agreed the specifications of a 25-seat glider in October 1940 and the initial production order for 440 aircraft to specification X26/40 was placed with Airspeed Limited as the Horsa.[168] By January 1941 jigs were being produced ready for manufacture and a mock-up conference had been held, which resulted in very few modifications to the initial design.[169] Flight trials commenced in August 1941, and by June 1942 the Horsa had been cleared for daylight troop-carrying operations, with further trials being conducted to clear it for night and poor weather flying.[170] The Horsa became the mainstay of the British glider fleet. It proved to be robust and capable on operations and was respected by those that operated it.[171] It only underwent one major modification during its life, where the cockpit and nose became hinged to enable faster offloading of equipment. The relatively unproblematic development of the Horsa was a remarkable achievement considering the scale of the technical challenge. However the risks of ordering straight from the drawing board were recognised and the development of a second specification, X25/40, for a

Equipment and Technology 83

15-seat glider was progressed alongside the Horsa for insurance. Slingsby Sailplanes Limited were handed the 15-seat specification, which resulted in the Hengist. Partly because of the success of the Horsa and partly because of structural problems, the Hengist, which incorporated many novel design features, was only produced on a very small scale and was never flown operationally.

The final British glider represented an immense technical problem and hence had a more complicated gestation. It was accepted that airborne forces would be vulnerable on the ground without support equipment such as vehicles, artillery, anti-tank weapons and preferably armoured support. It was also clear that, despite experiments in parachute delivery continuing throughout the war, the most efficient means of bringing this equipment to the battlefield was by glider. In October 1940 the War Office decided that the heavy glider should be capable of carrying an eight-and-a-half-ton tank (the Mark VII Tetrarch) or equivalent loads. It was subsequently suggested that rather than being designed to carry a specific vehicle, space for a generic load with limits from five to nine tons might prove more efficient. In either case the War Office emphasised the urgency of the requirement and urged the Air Ministry to do everything in its power to reduce the period of experiment and trials.[172] Another viewpoint was that the carriage of armoured vehicles was not as essential as being able to deliver transport vehicles and supporting weapons to the battlefield. If this was accepted then the Horsa might be adequate and the technically challenging larger glider unnecessary.[173] Lieutenant Colonel Rock, in command of airlanding development at CLE, emphasised the advantages if loads could be restricted to a size and weight that could be lifted by the Horsa. He argued that the tank-carrying glider represented a reduction in tactical flexibility, in that its size, weight and landing run would restrict it, and therefore the rest of the airborne force, to very few suitable landing zones. Support equipment should be adapted and developed to fit the Horsa rather than a new glider being produced to cater for current equipment. As has been shown, the technical challenges in producing support equipment were relatively minor compared to designing new gliders.[174] By mid-1941 a compromise had been reached. The tank-carrying glider would be produced but its maximum load was reduced to seven tons, enough to carry one Mark VII tank or two Bren Carriers or an equivalent load. In parallel to the development of this new model the specification of the Horsa would also be reviewed in order to ascertain if it could be adapted to carry a Bren Carrier, scout car or 15-cwt truck.[175]

The new tank carrier specification, X 27/40, was delivered to the General Aircraft Company for manufacture as the Hamilcar. By July 1941 a mock-up had been inspected and prototypes were under construction. Flight tests took place throughout 1942. The production of the Hamilcar was even more remarkable than the Horsa for its relatively trouble free development. The Hamilcar was an astounding aircraft; a glider weighing over sixteen tonnes with a useful load capacity of eight tonnes, it was over twenty metres long, had a wing span of thirty-four metres with the pilot sitting over seven metres above the ground, dimensions comparable to the Halifax.[176] The Hamilcar did not really achieve significant impact delivering tanks to the battlefield as originally envisaged. It did however prove invaluable in bringing heavy loads to the landing zones such as Bren Carriers, 17-pounder anti-tank guns and limbers, bulldozers and stores.

Despite the unavoidable policy of ordering direct from the drawing board the design and production of these bespoke glider types progressed remarkable smoothly. The Hotspur was a sound training aircraft and the Horsa and Hamilcar were highly effective operationally. Only the Hengist could be considered a failure and that aircraft had only been ordered as an insurance policy. However, all British gliders suffered from a common design fault that had not been fully anticipated and was to prove a major problem. They were all manufactured almost entirely from wood with a plywood skin covering a robust wooden frame. This policy was positively encouraged from the beginning as it would not interfere with mainstream aircraft construction and would therefore not compete with the Air Ministry's priority aircraft for resources. The furniture industry, a manufacturing sector previously without a direct war role, could be used for production.[177] This all-wood method of construction resulted in aircraft that, while physically robust, could not readily be broken down and reassembled for the purposes of transport. The wings could be removed and refitted by expert technicians, but this still left a hull the same size as a medium bomber to fit onto a truck or into a ship. The effort required provoked a deliberate policy that gliders would not be transported by road.[178] This did not present a problem when gliders had to be delivered between airfields in Britain as they could be towed by air. However, transport overseas presented a greater challenge as the aircraft took up too much space to be economical to move by ship. There was therefore no option but to fly them to foreign theatres, but the effort required to do this was incredible. Between 3 June and 7 July 1943 thirty-one Horsas were towed from Britain to Sale in Morocco under operation BEGGAR. The operation required eleven weeks training and during the 1,400 mile journey one glider went missing, two ditched in the sea and one had to return to Britain. During the onward transit to Kairouan in Tunisia another seven gliders had to be released over the desert due to technical difficulties. BEGGAR cost the lives of seven glider pilots.[179]

Clearly this rate of attrition was unacceptable. BEGGAR was conducted in order to build up a glider force for HUSKY; however more gliders were required for the operation than could be efficiently delivered in this manner. The answer came in the shape of the American Waco CG4A glider, henceforth referred to by its British name, the Hadrian. The Hadrian was built from a tubular steel frame with a canvas skin. This meant that it could readily be unbolted, broken down and crated for transport, and could then be rebuilt under supervision by unskilled labour. By 13 June 1943, 346 Hadrians had been delivered to North Africa. Of the 163 gliders flown by British glider pilots during HUSKY, 145 of them were Hadrians.[180] This in itself had serious repercussions. The Hadrian had a small payload of less than three-and-a-half tonnes, which meant it could only carry a jeep or an artillery piece, or artillery ammunition, and not the entire combination as the Horsa could. This presented obvious tactical disadvantages. Further to this the Hadrian, due to a different system of flight controls, had different flight characteristics to British gliders. In particular it had a shallower glide angle which made a precise landing spot much more difficult to identify and then achieve.[181] This was considered a distinct handicap when landing on unprepared ground and was made worse at night. In Sicily where the landing zones were small and bounded by thick stone walls, this characteristic, coupled with the Hadrian's less robust construction caused a high casualty rate amongst glider pilots and their passengers.[182]

The delivery by air of British built gliders to the Far East was obviously not practical. It was recognised that if India were to achieve its aspiration to operate its own airborne forces it would have to rely on the indigenous manufacturing industry to produce gliders. Considerable staff effort was expended on this issue and it was proposed that the furniture industry and railway wagon manufactures could provide the necessary industrial base.[183] This proved not to be viable due to the lack of high quality timber required to build the Horsa and Hamilcar. Production in India would entail importing timber and this was evidently uneconomic.[184] By August 1942 any plan to manufacture gliders in India had been officially halted.[185] After this decision the only recourse was to import Hadrians from America. It was estimated that over 2,000 Hadrians would be required in 1944 and 1945, however the only glider operation in the Far East involving British troops, Major General Orde Wingate's operation THURSDAY, used less than 100 gliders in total.

While the technical and engineering challenges associated with glider production were dealt with relatively efficiently the question of the quantity required was less simple to answer. The two subjects were, of course, mutually dependent. The important factor was the final quantity of troops and equipment that could be delivered to the battlefield. That overall load was the result of a calculation of individual glider capacity, depending on type, multiplied by the overall number of gliders available. However, this apparently simple process was complicated by a number of issues. Initial difficulty and disparity were caused by this simple equation either not being employed or by basing the calculation on uncertain assumptions. The CLE's initial attempts to estimate the number of gliders required for a brigade operation varied between 180 and 380.[359] Despite several attempts to arrive at a more scientific figure the projected total required to lift a brigade continued to fluctuate up until early 1944. The process was made more difficult by the equivocation over the operational status of the Hotspur as well as constant amendments to airborne formation establishments by the War Office. DMC expended considerable staff effort in order to create barely interpretable equations in an effort to produce reliable figures.[187] HQ AC Comd concluded that an accurate calculation was not worth the effort required to prepare it and that it was better to estimate a relatively safe figure and endorse it rather than to prolong ambiguity.[188]

A second area of uncertainty concerned the proportion of gliders required above the number necessary for a formation to conduct a single operation. This was important in order to inform manufacturers of the rate and target for production. Initially a figure of 10 per cent was assessed as sufficient to allow for wastage and training.[189] This was clearly a severe underestimate and was subsequently revised. Although glider recovery units were established, gliders used on operations had to be considered consumable and therefore each repeat operation would need a full complement to be manufactured. On top of this each operation would require a reserve to cover gliders damaged during transit or assembly. The training establishment also had to be provisioned, including a percentage for wastage through training accidents and wear and tear. This calculation resulted in a figure 250 per cent higher than the initial number of gliders required for a single operation.[190] Once the decision was made that planning should be based on three full-scale airborne operations being executed per year it became obvious that the annual production requirement for gliders was going to be immense.[191]

The practicalities of production could not wait for the War Office to arrive at a final figure for glider numbers. If glider production was to reach any useful level, design and assembly had to be started without delay. At the beginning of 1941 VCAS was instructed by CAS to place an order with MAP as soon as possible. This was months before any sort of rational estimate had been made and therefore any figure proposed by VCAS would represent little more than an informed guess based on early War Office predictions.[192] Nevertheless the order was placed and schedules for monthly production were developed based on this supposition. The Air Ministry was content with this approach as it represented a minimum diversion of staff effort in accordance with the priority it attached to airborne forces in general. MAP was also initially happy to proceed on this basis, but six months later the Minister, John Moore-Brabazon was concerned that the requirement had not been sufficiently refined. He was content that the initial order placed by VCAS would be met and therefore the War Office's immediate needs could be satisfied. However, it was clear to MAP that this only represented the initial order and that unless continuation orders were placed quickly production would come to a halt in mid-1942, with only a single brigade's lift having been manufactured. Unfortunately, with a firm and final estimate of numbers still not available the Air Ministry felt unable to provide anything other than a partial reply to MAP's request which, by its own admission, amounted to no more than 'helpful directions.'[193] The Air Ministry was not going to volunteer a request for increased production to MAP if it did not have to. While glider production did not directly challenge the Air Ministry's priorities for resources it did represent an indirect threat, as each glider required an aircraft to tow it. The scale of that indirect threat was now apparent, and continued ambiguity in the requirement therefore suited the Air Staff's purposes, an example of passive resistance through inactivity.

A lack of useful engagement from the Air Ministry meant that any shortfall in glider production went largely unchecked. Early in 1942 Browning was uncertain that more than a fraction of the production target would be met and that this would affect the progress and morale of the Airborne Forces Establishment (AFE). The Air Ministry agreed with Browning, confessing that it also placed little faith in the output predictions, but did little to refute MAP's assurances that the production rate had reached its ceiling.[194] The situation changed when Browning gained direct access to MAP via Renwick's committee and, although production continued to lag behind schedule, by May 1942 the War Office was content that its requirements would be met by March 1943.[195]

Considering this accepted lag in production it must have come as a surprise when Sir Stafford Cripps wrote to Churchill later in 1942 suggesting that production was exceeding the point at which the number of gliders being manufactured could be usefully employed. He suggested to the Prime Minister that consideration should be given to discontinuing the programme altogether.[196] This came at the point when the entire question of the future of airborne forces was being referred to the Prime Minister for arbitration and the Air Ministry co-opted Cripps' views as part of their argument for a reduction in airborne development. Lord Cherwell, Churchill's chief scientific advisor, also concluded that a substantial reduction of the glider programme should be considered.[197] The Prime Minister acquiesced and ordered an immediate curtailment of Horsa production.[198]

Fortunately for British airborne forces, SYMBOL and the decision to begin planning for HUSKY countermanded Churchill's original decree. Any temporary recession in glider numbers due to his curtailment order was compensated by the influx of American gliders into North Africa in preparation for the invasion of Sicily. With a lull in airborne operations between HUSKY and OVERLORD glider numbers recovered to the extent that, by the end of 1943, the War Office felt confident enough to offer the

Americans an allotment of Horsas and Hamilcars for their training in Britain.[199] By the time OVERLORD was imminent the War Office had learnt a lesson and kept a close eye on production levels through an inter-departmental glider committee, an offshoot of Renwick's committee. Six divisional level airborne operations were predicted for 1945 and gliders were ordered to meet this requirement. Even after D-Day the War Office warned against curbing the programme without very careful consideration.[200]

The British glider programme between 1940 and 1945 represents a remarkable success. The technical problems were immense and the engineering challenges were also considerable. Despite this new aircraft types were developed from scratch, a feat accomplished with very few failures. Working out the number of gliders required was less straightforward but this is understandable considering the constantly fluctuating baseline on which the estimates were built. The General Staff was continually refining the order of battle of airlanding units, while more extensive adjustments to the establishments were forced by the War Office and Air Ministry's divergence over the size, shape and future of Britain's airborne capability. Less explicable is how a perceived deficit in numbers was so quickly transformed into an alleged surplus, which in turn influenced Churchill in making his decision to significantly reduce airborne development, albeit temporarily. Certainly the Air Ministry, concerned by an indirect threat to its resources, was reluctant to engage closely with the problems associated with glider production and this did not assist. It did little to support the CLE and AFE to formulate a workable solution nor did it help MAP to understand the nature of the problem. The assessment that any Air Ministry enthusiasm on paper did not translate into practical assistance is probably accurate.[201]

The most significant fact concerning glider production is that the challenges that arose appear not to have seriously impinged on airlanding training and operations. This is despite many factors that threatened glider production throughout the war. The adequate supply of plywood was a persistent problem, and the Air Ministry frequently sought to give priority in supply to the De Havilland Mosquito and training aircraft types over glider manufacture.[202] Other potential engineering threats to production included a shortage of draughtsmen and a lack of suitable storage facilities for gliders coming off the production line. The War Office also caused difficulties by appearing to be unable to produce a list of the standard loads to be carried by glider. It constantly amended and added to the inventory as new equipment became available and the organisation expanded. This had impact on the CLE, which either expended nugatory effort on trials for equipment no longer required or was forced to wait before equipment was confirmed and trials could be conducted.[203]

However, the main threat came from the lack of engagement from the Air Ministry and its attempt to protect its resources through passive resistance. Nevertheless there is no evidence in the various published memoirs of glider pilots that there was any significant shortage of gliders at any stage, except in the very early stages of development and briefly in North Africa prior to HUSKY. Glider pilots achieved high rates of flying during training and 200 'lifts' per pilot per year was considered to be only slightly above average.[204] Brigadier George Chatterton, the Commander Glider Pilots, confirms that it was not a deficit of gliders that caused training problems. In fact the opposite seems to have been true when he commented, 'a curious situation was arising with regard to the Horsa gliders. Many were coming from the factories, hundreds in fact, and they were parked all around RAF Netheravon, but there were hardly any tugs that could be spared to tow them into the sky.'[205] Chatterton's observation appears to support Stafford Cripps' and Cherwell's assertion that gliders were being produced in surplus quantities. It also demonstrates the relative impact on training of the availability of support aircraft versus the provision of gliders.

Procurement and Supply as a factor in Development.
The availability of suitable aircraft both to drop paratroops and to tow gliders was the dominant equipment factor that affected the development of Britain's airborne forces, far more so than the provision of personal and support equipment or the procurement and supply of gliders. Shortfalls in aircraft availability impinged to some degree on individual and collective training and exercises, which were consequently sometimes constrained in scale and scope.[206] Whether that impact on training had any significant effect on the overall progress of airborne development and hence military effectiveness will be assessed in the following chapter. However, it is already possible to see how this line of development was an immediate factor in military effectiveness. A deficiency of aircraft directly reduced the size of an airborne force that could be committed to an operation. If this was unacceptable then aircraft had to be provided from other sources. This substitution of aircraft by the USAAF and Troop Carrier Command had significant operational consequences during TORCH and HUSKY. Following TORCH, CIGS implored the COS Committee to accelerate the provision of transport aircraft by every available means, and Alexander clearly identified similar lessons post HUSKY.[207] Cooperation did improve between American aircrew and British airborne troops, through compulsion as much as any other factor, as American aircraft were required to lift British airborne forces in all the major airborne operations of the war. The only exception was D-Day, where the short flight distance allowed RAF crews to fly multiple sorties in the same day between England and Normandy. Aircraft shortages also caused commanders to adapt their tactical doctrine to fit the availability of lift, which then had an effect on the manner in which an operation was conducted. This subject will be examined in depth later.

It is worth considering why the availability of transport and support aircraft proved to be such a chronic problem throughout the development process. Financial constraints can be disregarded as a significant factor as the Cabinet had already decided as early as April 1938 that the scale of re-equipping the RAF would not be defined in financial terms but would be dictated solely by industrial capacity.[208] Barnett and Postan's assertions that the British aircraft industry had an innate production ceiling caused by industrial torpor and other unpredictable and intangible influences can only be regarded as a backdrop to this specific problem. This situation would have created equal drag across aircraft production whereas in fact the provision of transport aircraft and the modification of bombers for airborne forces were more acutely inhibited than other areas of the programme. It could also be assumed that the relative levels of technical and engineering effort required to provide suitable support aircraft was an underlying factor. As a comparison it is undeniably true that the technical and engineering endeavour required to supply personal and support equipment to airborne troops on the ground was relatively low and this led to its efficient provision. However it can also be argued that the technical and engineering challenges associated with manufacturing gliders were greater than those required to modify bombers for airborne operations or even to produce a new model of transport aircraft. Despite this the provision of gliders was less problematic than the availability of support aircraft.

The actual reasons behind the situation can be deduced by comparing who owned the requirement for the equipment against where control of the resources necessary to fulfil it was invested. The requirement to acquire the entire range of equipment necessary to support airborne forces was owned by the War Office. In the case of personal and ground support equipment the General Staff also principally controlled the resources, via the Ministry of Supply, to fulfil that requirement. This meant that the War Office could allocate the priority that it saw fit to the provision of this equipment. It could also plan its own acquisition strategy, such as adapting and modifying equipment already in service or collaborating with the US. Postan asserts that the provision of this class of equipment was an added demanding burden on the Ministry of Supply but in fact the relatively small scale and low complexity of ground equipment for airborne forces resulted in the requirement being met with comparative ease.[209] In contrast, control of the resources required to supply gliders was exercised by the Air Ministry via MAP. However the use of wood in their construction and the unconventional manufacturing base meant that the provision of gliders did not directly compete with other Air Ministry programmes and hence it was initially allowed to proceed unhindered. It was only when the scale of the glider force became apparent that the Air Staff became concerned about the indirect impact that it would have on its resources, that is the use of bombers to tow them. By that time most of the technical and engineering challenges had been overcome and it was only the question of quantity that became an issue.

The Air Ministry and MAP also had control of the resources required to provide support aircraft but in that case there was a real and direct threat to the Air Staff's priority programmes. In the early part of the war precedence was given to the manufacture of fighter aircraft and Beaverbrook strenuously opposed any diversions. As the threat of invasion subsided the main production effort began to switch to the bomber force. There had been great difficulties in the development and initial production of the modern bombers which were obstinately slow to reach the front line throughout 1941, and with only 85 per cent of the projected force available at the end of that year the RAF was loath to commit them to any activity outside its core doctrine.[210] With control of the resources the Air Staff, through MAP, were able to passively resist the parachute and glider modification programmes by simply not allocating them adequate labour, materiel and staff effort. There was very little the War Office could do to expedite the process. Transport aircraft therefore never achieved a priority that allowed them to compete for production resources with bomber and fighter aircraft. The Air Staff could easily resist War Office demands by stating the false assumption that British requirements could be fulfilled by American production. By the time it was apparent that this claim could not be achieved it was too late to effectively recover the situation, with the results that have been described.

Finally, the Air Staff actively resisted opportunities to provide transport aircraft to support airborne forces. When General Staff members of Renwick's committee visited the US in May 1944 they were impressed with the potential of two new transport aircraft being produced by the Americans: the Fairchild C-82 and the Budd (Conestoga) C-93. The Director of Air from the War Office was therefore dismayed to discover on his return to Britain that the C-93 had been offered on Lend Lease several months before but had been turned down by the Air Ministry without reference to the General Staff.[211] Even when transport aircraft did become available the War Office could do little to influence where the Air Ministry allocated these resources, hence the first Dakotas that became available in Britain were assigned to mail delivery rather than to supporting the General Staff's requirements. The misalignment of the requirement for equipment with the authority to control the resources to fulfil it was the critical factor that created the chronic shortage of aircraft to support British airborne forces.

However, the Air Ministry cannot be directly blamed for adopting a less than fully committed approach to airborne development. It had other priorities to pursue and before 1943 it could perceive no obvious imperative to alter its stance. This situation was perpetuated by the failure to create any joint establishment with the power to influence and align the requirements and effort of both the War Office and the Air Ministry with respect to airborne development. Without this the priority placed on development was disjointed and situations such as the paucity of transport aircraft were inevitable.

CHAPTER FOUR

Personnel and Training

Manpower

Manpower was one of the major resources required to generate British airborne forces during the war. Human resources in Britain were at a premium and never less so than during the period of rapid expansion during the first half of the war. Between June 1939 and June 1943 the armed forces grew by nearly 800 per cent from 480,000 to 4,300,000 men. The needs of the armed forces, the national services and labour supply were not effectively coordinated until a Director-General of Manpower was appointed within the Ministry of Labour in August 1941.[1] Prior to this date the Army and other services were at liberty to recruit towards the projected size of their forces, as they considered necessary. Up until the outbreak of war the Army was losing the recruiting battle, with men preferring to go to the Royal Navy and RAF where pay and prospects were better.[2] Every Army recruit was therefore valuable as he represented a minor victory over the other services and the Ministry of Labour. It is perhaps unsurprising then that the War Office took exception to Churchill's frequent requests to give up manpower to provide the resources for his many irregular schemes and ideas, the merits of which were often considered dubious at best. For example the Prime Minister, against opposition, had personally to fight through the formation of the first commando units.[3]

Early in 1941, in order to halt its spiralling demands the Prime Minister proposed a net entitlement for manpower above which the Army would not be permitted to recruit. Churchill ordered the War Office to make best use of this entitlement 'by wise economies, by thrifty and ingenious use of manpower, by altering establishments to fit resources.' The cap was set in March 1941 at 2,195,000 and, although this figure was increased to 2,374,800 by autumn 1941, this represented an absolute ceiling to the establishment of the Army. Any increases due to new formations being raised from that point forwards, such as airborne forces, would have to be met through internal recruiting and re-appointing existing units.[4] Lateral thinking and pragmatism would have to be exercised by the General Staff in order to meet the airborne forces manpower requirement.

Manpower implies a raw resource and therefore training, both individual and collective, would also be required in order to produce an effective physical component. The training of airborne forces, principally paratroops and glider pilots, posed several challenges. Obviously training had to precede any thought of operations; however Britain had no experience on which to base this training. It has been suggested that the British Army at home between 1940 and 1944 trained on the basis of the experience being gained in the Mediterranean and therefore it did not have to rely on theoretical precepts.[5] However these conditions were not applicable to airborne forces. There were no major British parachute operations until the end of 1942 and no full airborne operations until mid-1943. Therefore training, particularly collective training, was a theoretical pursuit for the first half of the war. Without indigenous experience the early trainers would have to look elsewhere. The War Office investigation into other nations' airborne capability in the mid-1930s was not conducted in sufficient depth to reveal their detailed training methods. An article translated from the French 'La Revue d'Infanterie' in 1936 may have given the War Office some insight into 'a new subject and how engrossing for your officers!' The French had identified the requirement for tactical training, collective training and technical parachute training.[6] A report submitted by the Air Attaché in Berlin in 1938 provided information that had appeared in the German press. This covered in brief detail the parachute training process, including the objectives of ground training and the process of a training jump from an aircraft.[7] However this short-lived period of information gathering, instigated in part by Dill, was left largely ignored at the beginning of the war. Airborne training was an unknown in 1940 and it would require a degree of vision and considerable ingenuity with both processes and systems in order to develop a successful comprehensive training programme.

The Recruitment and Selection of Paratroops
In a minute to Ismay on 3 June 1940 Churchill ordered the formation of geographically discrete, well-equipped, self-contained commando units of approximately 1,000 men each.[8] Less than three weeks later the War Office was instructed to find another 5,000 men for the Prime Minister's parachute force. The General Staff took the pragmatic, although perhaps slightly disingenuous decision to combine the two requirements. By 23 June 1940 twelve commandos had been designated for establishment, two each under the control of Southern, Eastern, Western and Northern Commands and

one each in Scottish Command and Northern Ireland District. The remaining two commandos were to be under direct control of the War Office. No. 1 Commando had already been formed from personnel in existing Independent Companies and was based in Southampton. No. 2 Commando was designated as the parachute commando to be based at Ringway near Manchester. No. 2 Commando was to be formed as a priority from an amalgamation of commando troops recruited by the various commands across the country.[8]

On formation, No. 2 Commando, despite its unique role, was subject to the policies and regulations governing all of the irregular commandos. The initial paratroopers were therefore recruited against the standards required of all commando troops. These were not particularly specific. In general all the men were to be volunteers: young and absolutely fit. In addition officers were expected to have tactical ability and imagination and soldiers were to display a good level of independence and total reliability.[10] Additional, more specific prerequisites were subsequently applied to volunteers for No. 2 Commando to ensure they were suitable for parachute duties. These were a maximum personal weight when clothed and lightly equipped of not more than 250 pounds, the ability to pass comfortably through a circular aperture of 3 feet diameter with equipment and an absence of physical disabilities such as thin skulls and weak ankles.[11]

Before parachute training could begin it was recognised that a programme of basic tactical training would be required. Commando volunteers came from every arm of the Army, including cavalry, infantry, Royal Artillery (RA), Royal Engineers (RE) and Royal Army Service Corps (RASC). Hence standards and experience would be different depending upon the origin of the volunteer. Therefore to standardise the skills across the Commando basic weapon and tactical training was considered essential. Physical fitness was also emphasised in order to equip the commandos to cope with the harshness of raiding operations.[12] Aside from these physical requirements there were also a number of psychological pressures that were unique to airborne operations and would have to be addressed in order to produce effective paratroops. First there was the obvious fact that parachuting was an unnatural process that produced strong natural reactions against it.[13] Second was the mental dislocation caused by an airborne soldier departing his domestic station in England and dropping into the heat of battle just a few hours later. Frost observed this phenomenon. 'We had not appreciated the effects of the sudden transition to which the airborne soldier may be subjected. From comfortable billets with beds, clean clothes, hot water, extra rations and a very fair share of drink [the airborne soldier] then finds himself struggling over hostile territory with no creature comforts at hand.'[14] Lieutenant Colonel Napier Crookenden, the commanding officer of 9 Parachute Battalion, recorded on the eve of Operation VARSITY, 'One of the hardest things to realise as I sit here in the English sunshine is the fact of our translation in a few hours in both a different country and a very different environment.'[15] The effect of these two psychological stresses and attempts to mitigate them will be examined later.

A third form of mental pressure might also be experienced by paratroopers who became separated from their comrades during an operational parachute drop and found themselves isolated on the battlefield. British doctrine attempted to define this phenomenon. 'The qualities of individual resource and initiative are desirable in all fighting troops, but are particularly necessary to parachute troops in the period between dropping and reaching the unit RV [rendezvous], when they may be deprived of the leadership of their officers and NCOs and the support of their comrades.'[16] Reduced to an individual with no group for reassurance this experience, particularly at night, had the potential to paralyse a soldier. Many paratroops referred to this reaction. 'I must admit when I hit the ground and looked around, I thought I was the only bugger there! It was so quiet.... There was a deadly silence; all I could hear was the wind blowing through the grass. This was the moment I was really scared.'[17] Many paratroopers describe 'the feeling of horror and the rapid draining away of self-confidence....'[18] British doctrine recognised that at this moment paratroopers were particularly vulnerable. 'The most critical period for airborne troops is the few minutes immediately after landing.... Taking into consideration also the nervous strain imposed on parachute troops immediately before they descend, it follows that their morale in the early stages [of an airborne landing] is likely to be very low.'[19] Steps had to be taken to ensure that paratroopers were mentally equipped to cope with the pressure of this stage of the battle. Personal independence was identified as the key factor. This went against the prevalent attitude towards training, which taught soldiers that their first duty was instant, unhesitating and exact obedience to their superior's orders.[20] However, airborne soldiers had to be selected or trained to act under their own initiative until they could meet up with their comrades and become an effective team member.

The generic selection criteria for commando personnel stressed the requirement for a good level of independence and total reliability. It was initially believed that this could be achieved by selecting 'tough' men. The commander of No. 1 Parachute Training School (PTS) confirmed, 'the standard of men accepted [for service in No. 2 Commando] was indeed 'tough'.'[21] However, this policy was exposed during an inspection of Ringway in August 1940 as not necessarily producing the correct calibre and quality of volunteer. The inspecting officer recorded his concern.

> I am of the opinion that far too many of the men looked 'tough', but not mentally alert and it is suggested that the type required is rather of the

dirt-track racing, professional footballer, or rugger player type who presumably have both toughness and mental alertness combined with willingness to accept risks and go hard, rather than the man who is tough because he has not the mental equipment to think of the consequences of his actions.[22]

Trying to inculcate personal independence through the men's living conditions was also unsuccessful. The commandos, including those in No. 2 Commando, were not provided with food or accommodation. Instead they were paid a daily allowance of six shillings and eight pence in order to cover lodging, feeding and transport. Looking after themselves was supposed to encourage independence and initiative. However it was found that living away from a military environment encouraged late nights and was therefore detrimental to physical fitness. The system was difficult to administer and it was considered doubtful whether independence and initiative really were improved to any great extent.[23]

The only sure way to improve a soldier's ability to act effectively when isolated was through training. This had in fact been identified prior to the war in order to equip men to cope with dispersal on a modern battlefield.[24] Even during the brief surge of staff activity connected to airborne forces during the 1930s it was recognised that particular training techniques were required to produce men with the right attitude. Periods of 'hard living' were included in projected training schedules.[25] With the formal establishment of parachute training in July 1940 mental fortitude was linked to physical fitness. Parachute troops were to be kept at the highest pitch of physical fitness in order to withstand the mental as well as the physical rigours of airborne operations. Provision was made in the syllabus for intensive physical training instruction.[26] When 1 Parachute Brigade was established in September 1941 the headquarters was located at Hardwick Hall near Chesterfield. A formal programme of ground training was introduced at the new centre for recruits prior to parachute training and for the trained paratroops of the expanding brigade. Hard physical and tactical training was used to instil a sense of confidence, which was intended to assist in overcoming the psychological pressure of isolation on landing. The syllabus was designed to test officers and soldiers to the very limits of their physical endurance, and trainees recount the mental effort required to prevent themselves from giving up.[27] Despite this some individuals such as Crookenden appear to have enjoyed the experience. 'It will be a pleasant change from my present work, as I shall spend 10 days on the equivalent of a P.T. course – No thinking, but plenty of action.'[28]

This approach to training appears to have achieved its aim based on the evidence of individual paratroopers' conduct immediately on landing during operations, particularly during OVERLORD where the drop was at night and scattered. Many soldiers who were temporarily isolated reported feeling apprehension or even fear but all appear to have been able to overcome their base instincts and make efforts towards achieving their mission, such as navigating to a rendezvous or reacting to enemy activity in the most appropriate manner. Some managed to survive and evade the enemy for days on their own initiative, having inadvertently landed many miles from their DZ. Brigadier Poett, Commander 5 Parachute Brigade, recalled his reaction after landing in Normandy. 'I had no idea where I was. It was too dark to see the church or any of the landmarks on which we had been briefed, but I could see the exhaust of the aircraft disappearing and I knew that it would be going over Ranville. I knew my direction therefore… I moved in the direction of flight of my aircraft, and sure enough I came across one of my men.'[29]

In mid-1941 the War Office ordered an expansion of the airborne establishment from a single battalion to a complete brigade. In doing so the General Staff adhered to its policy of all paratroopers being volunteers.[30] No. 2 Commando, which had been renamed 11 Special Air Service (SAS) Battalion in October 1940, formed the nucleus of 1 Parachute Brigade and was renamed yet again as 1 Parachute Battalion. The remaining battalions would be formed as new units by extracting volunteer manpower from across the Army. When the request for volunteers from all British infantry resources in the United Kingdom was promulgated in August 1941 a limit of a maximum of ten other ranks volunteering per unit was set so as not to deplete any one unit unduly.[31]

Even with this limit imposed the policy of recruiting volunteers from the infantry in particular created resentment in many commanders. The difficulty was not necessarily the quantity of volunteers required but the quality of the soldiers that it drew away from other units. General Sir Brian Horrocks described the problem.

> It would be safe to say that out of a section of, say, ten men, two lead, seven are perfectly prepared to follow where they are led, and one would much prefer not to be there at all…. One of the reasons why so many generals objected to men being asked to volunteer for special cloak-and-dagger private armies was it was always the leaders who volunteered. In these special formations… each leader represented only himself as they were all of the same type; but in his regiment he was worth almost a whole section, for he was the man the others would follow.[32]

Gale was certain that the exceptionally high physical and mental standards required of paratroopers necessitated a volunteer intake and was worth this criticism from the rest of the Army.[33] Certainly later in the war there is evidence that calls for recruits for airborne forces were obstructed through such methods as deliberately not circulating the Army Council Instructions calling for volunteers down to unit level.[34]

Although training had been recognised as the only sure way of producing a paratrooper with the necessary qualities of independence and initiative, a selection process was maintained. More precise physical standards were drawn up. Prospective paratroops had to be between nineteen and forty years old, not over six feet two inches tall and weighing no more than 182 pounds. Visual acuity had to be not below 6/12 in each eye without correction and hearing had to be up to Hearing Standard 2.[35] There is not much evidence for selection procedures beyond a medical to confirm physical standards. In June 1941 the Adjutant General, General Sir Ronald Adam, recommended to the Executive Committee of the Army Council that psychological testing should be introduced in order to increase the efficiency and effectiveness of recruit selection across the Army.[36] Gale confirms that the introduction of psychological testing for potential parachute recruits reduced the wastage rate during training.[37] Although no primary evidence appears to exist concerning this selection procedure it is probable that it was part of Adams's wider reforms rather than any unique innovation on the part of airborne forces.

The reasons for a man volunteering to become a paratrooper were many and varied, some altruistic and others more basic. Undoubtedly there were some who were not altogether aware of exactly what they were volunteering for. Both Frost and Brigadier Alistair Pearson recall, as junior officers, not knowing that the 'special service' they had volunteered for involved parachuting until they were already most of the way through the selection process.[38] Still others knew exactly what they were doing and, particularly during the early part of the war, volunteered as a means to take the battle to the enemy more quickly than might otherwise have been the case.[39] Some volunteered because the glamour appealed to them and others just to escape their current situation.[40] During late 1943 and early 1944 men serving in the Middle East applied to secure a posting back in Britain.[41] The various badges and symbols enhanced the perceived glamour of airborne forces. The maroon beret and the badge depicting Bellerophon astride a winged Pegasus in sky blue on a maroon background were chosen personally by Major General Frederick Browning in order to foster the principles of high morale and a *corps d'élite*.[42] The qualified paratrooper's badge (a pair of blue wings with a parachute in the centre) and the maroon beret in particular quickly became synonymous with British airborne forces, and while the beret may not have directly induced men to volunteer there is evidence that it became a coveted item by those already considering volunteering. Crookenden, on being posted to 6 Airborne Division after Staff College in 1943 wrote to his parents commenting, 'I look very swell in my purple [sic] hat.'[43]

Money was certainly a motivating factor. The normal pay of soldiers was a constant cause of irritation and criticism. The War Office Committee on Morale in the Army reported that of a sample of the negative letters received by the British media during 1942 and 1943, 59 per cent were complaints over pay and allowances.[44] The promise of extra pay would have been a strong attraction to many soldiers. When the independent living experiment of No. 2 Commando failed the subsistence allowance was withdrawn and replaced by parachute pay. The rate was set at four shillings per day for officers and two shillings per day for other ranks. The commanding officer of 11 SAS Battalion questioned this disparity and suggested that the two rates should be made equal at three shillings per day but there is no evidence that this was done.[45] This did not appear to cause problems as one soldier recalled his comrades in training as, 'volunteers, for many of whom no doubt the extra 2/- per day when qualified was a powerful incentive to succeed.'[46]

The role of propaganda has also been cited as one reason for soldiers volunteering for service as paratroopers.[47] Propaganda in the shape of a recruiting campaign was required in the early months and years of the war to advertise what was a virtually unknown form of soldiering. Later it was needed to keep airborne forces, whose operations were few and far between, in the public eye in order to maintain a steady stream of recruits. The propaganda took two forms, direct and indirect. Direct propaganda was official War Office material, aimed at serving soldiers via the chain of command to persuade men to volunteer for parachuting. Lecture tours were organised where parachute trained officers could describe their experiences and answer questions from units around the country.[48] However it would appear that the effect of the appearance of the lecturers on the audience was not always taken into consideration. As one prospective paratrooper recalls, 'Of the two officers who presided, the first, Brigadier Bill Gough [commander 50 Indian Parachute Brigade], had his leg in plaster, while his Brigade-Major was encased from hips to armpits. The brigadier announced, "We are scarcely good advertisements for parachuting: none the less we invite you to volunteer."'[49]

Official literature was also produced with the intention of non-airborne regimental officers using it to deliver lectures to their own soldiers in order to persuade them to volunteer. An example of this was a two part article entitled 'Parachuting as a Career', written by a War Office staff writer, Major Anthony Cotterell. The article appeared in the official periodical 'War' issues number fifty-nine and sixty in November and December 1941. The article was a detailed and humorous account of parachute training. At the end of the article there were a number of lines of thought and specific points for the presiding officer to introduce into any subsequent discussion. There was even the suggestion that the officer running the session might demonstrate a parachute landing roll, 'When done correctly it is infectious.'[50] The value of this approach must be considered dubious. There is more than a touch of the blind leading the blind and, bearing in mind some commanders' reluctance to release volunteers, it is probable that these internal lecture sessions were delivered with a lack of enthusiasm, if at all.

Indirect propaganda appears likely to have attracted more volunteers, reaching a wider audience and being presented in a more appealing format. This involved ensuring that suitably targeted articles on parachuting and airborne forces appeared regularly in national periodicals, publications and featured on national newsreels and in films where security allowed. In October 1941 Ringway organised a three-day media event during which press photographers, correspondents, journalists, filmmakers and representatives of the BBC were invited to attend. As a result, the following month Gaumont released a newsreel to cinemas across the country describing some of the training of paratroops, demonstrating a mock operation and ending with the stirring proclamation (if somewhat optimistic at the end of 1941), 'And that's what Britain's paratroops are doing. So look out, Mr Nazi; we'll be seeing you!'[51] Most magazine articles were equally fervent in keeping with the national mood, although the tone was often tailored to a specific audience. Therefore an article in Picture Post was detailed and lavishly illustrated with black and white photographs with the reporter's account of his own training and parachute jump experience. The Boy's Own Paper took a more adventurous approach while Flight Magazine contained much technical detail.[52] The potential impact of this indirect propaganda was such that in early 1942 an officer, Lieutenant Colonel W.A. Sinclair, was posted permanently to the central airborne staff to coordinate the publicity effort.[53] It is difficult to assess the success of all propaganda in persuading men to volunteer. References in memoirs confirming the authors or their colleagues being influenced by lectures, films or magazine articles are infrequent. Whatever the true influences that persuaded men to become paratroops, volunteers were initially forthcoming in sufficient numbers to fulfil the requirement up to the strength of a parachute brigade.

In 1942, however, Britain's airborne capability began to expand exponentially. In the spring the War Office decided to form a second parachute brigade in England to make up the formation of 1st Airborne Division, and in response 2 Parachute Brigade began to form on 17 July. In November 1942 orders were issued for the formation of 4 Parachute Brigade in the Middle East. On 23 April 1943 the War Office ordered the formation of 6th Airborne Division, which necessitated the establishment of another two parachute brigades through that year. The manpower requirements to meet this rapid expansion were unlikely to be met through a purely volunteer intake and a different approach had to be considered.[54] Detailed thought had been given to the methods of raising parachute battalions in mid-1941. One of the options discussed was to discard the volunteer principle and detail entire battalions to undergo parachute training, replacing those individuals that were found to be unfit. The advantages of this method were clear. It was the quickest way of forming a parachute battalion; it provided an administratively efficient solution and the hierarchy and *ésprit de corps* of the unit would be retained. However, disregarding the volunteer principle was an unknown quantity and the repercussions were considered possibly dangerous. Nevertheless, if speed of establishment was the main consideration then this option was recommended.[55] Rock was less concerned with the possible disadvantages of mass conversion of battalions to the parachute role. He was not sure whether a volunteer was any better than a pressed man as very few volunteers had any real conception of what they had let themselves in for anyway.[56]

In mid-1941 the decision had been taken to retain the principle of a volunteer-only force during the expansion to form 1 Parachute Brigade. From mid-1942 the situation had changed and, although no operational imperative yet necessitated speed of formation, the scale of expansion required the most administratively efficient process possible. Entire infantry battalions were allocated for conversion to parachute battalions. On initial examination it would appear that the potential results of this scheme could have been estimated, after all No. 2 Commando had been converted *en masse* to paratroops and 85 per cent of the men successfully made the transition to the parachute role. However, these were men who had already been selected for their character and physical attributes, unlike standard infantry battalions. The infantry units of 31 Independent Brigade Group had been converted to airlanding infantry from October 1941 to form 1 Airlanding Brigade, as were its supporting arms such as engineer squadrons and artillery batteries. However, the parachute role was more demanding than that of airlanding infantry so this process also did not provide a satisfactory model. Therefore, as had been identified in 1941, the result of converting complete battalions could not be predicted. Certainly taking No. 2 Commando's 85 per cent conversion rate would have been over optimistic. Lieutenant Colonel Pine-Coffin, the commanding officer of 7 Parachute Battalion, recalls a 70 per cent success rate when 10 Battalion, Somerset Light Infantry converted.[57] However, the primary evidence appears to question this figure as improbably high. When 10 Battalion, The Green Howards converted to 12 Parachute Battalion, only 14 out of 24 officers were successful and 147 out of 636 other ranks. This equates to approximately 25 per cent of the original battalion. Similarly, when 2/4 Battalion, The South Lancashire Regiment became 13 Parachute Battalion, only 13 out of 24 officers and 210 of 766 other ranks converted, a rate of less than 30 per cent.[58] These figures were offset to a degree by the reduced establishment of a parachute battalion compared to a regular infantry battalion. A parachute battalion consisted of approximately 450 men of all ranks. Therefore the post conversion figures would have equated to 35 and 50 per cent of the manpower required for 12 and 13

Parachute Battalions respectively. Even so there was still a significant shortfall that had to be made up by volunteers from other infantry battalions across the Army.

Despite this wastage rate and continued reliance on volunteers to provide the majority of paratroops this system did still have advantages. The figures show that over 50 per cent of a battalion's officers were successful at conversion. Also, of the percentage of other ranks that did convert a high proportion were the NCOs of the battalion. Therefore the integrity of the hierarchy and the unit identity were still retained.[59] This was an important fact as the command and control mechanism within a battalion remained intact, which could improve efficiency, particularly during training and operations. The process also allowed a unit to shed its undesirable individuals without the usual bureaucracy. One artillery officer considered that it was a privilege to be able to rid themselves of the personnel that they assessed as detrimental to unit effectiveness.[60] In many cases it is unlikely that the victims of this process were even aware of the reasons why they had failed to make the grade.[61] In retrospect this policy might be considered misguided, breaking up perfectly serviceable infantry battalions and still requiring a significant recruiting effort. However the operational record of those converted battalions is at least equal to those raised from a purely volunteer intake. Thus, this process of selecting a battalion for conversion and then back-filling with volunteers proved to be a satisfactory compromise between speed and efficiency of formation while still retaining the volunteer ethos.

Recruitment at the beginning of the development process was retarded to some extent by the all-volunteer policy and the rest of the Army's understandable reluctance to release high quality manpower to the new establishment. This suited the War Office's aspirations, which in the early part of the war were an eagerness to maintain the manpower requirement at a low level. In fact there would have been little benefit in increasing the rate of recruitment early in the development process and before the training establishment was in a position to cope with the throughput. From 1942 onwards the willingness to compromise the volunteer principle allowed flexibility in the recruitment process, which in turn facilitated rapid expansion while retaining the qualities that made parachute units unique. Less easy to identify are those factors that drove individuals to volunteer for such an unknown and dangerous profession, often in the face of resistance from their superiors. Certainly ignorance, pay, image and propaganda played their part but none of these factors could compel a man to volunteer against his own volition. Even after the losses suffered by 6th Airborne Division during Operation OVERLORD (June 1944) and by 1st Airborne Division during Operation MARKET GARDEN (September 1944) there was still a steady stream of volunteers willing to fill the vacancies.[62] However, as recruits were secured they still only represented a raw manpower figure. Specialist parachute training was required to produce effective paratroopers and the rate of training would have to keep pace with expansion if bottlenecks were to be avoided.

Parachute Training
Following Churchill's minute of 20 June 1940 parachute training facilities were established with remarkable speed. Before that date military parachuting in Britain was practically non-existent. There was a small parachute section at RAF Henlow but this was for the provision of life-saving parachutes for aircrew, not the deliberate insertion of soldiers. Manchester Corporation's civil airport at Ringway was chosen as the site for parachute training despite reservations, which will be examined later. There were no qualified parachute jump instructors available. Instead a handful of airmen, fabric workers at Henlow, who had formed a semi-official parachute display team, were drafted in to teach parachuting. These were joined by a few NCOs from the Army Physical Training Corps (APTC) who had volunteered for parachute instructing duties despite having no experience of the process themselves. By 8 July most of the initial instructors and staff were present at RAF Ringway and the first 100 men from No. 2 Commando arrived on the same day. The first live drops took place on 13 July.[63]

The pace at which these initial developments were made caused nervousness in some areas of the Air Ministry. Slessor expressed his disquiet as he often did over airborne forces.

> I am rather uneasy about the air side of the development of parachute troops, and am afraid if we are not careful that it will be a case of more haste less speed; I am also a bit afraid that if we try to go too fast we may have unnecessary training casualties which will be a set-back to the development of the parachute units.... I am very anxious not to prejudice the success of the new organisation by trying to rush it unduly at the beginning, especially as morale and confidence are such an important factor in this respect.[64]

On this occasion Slessor's fears were well founded. Minor injuries were apparent from the very first drop at No. 1 Parachute Training School (PTS), Ringway, and on 25 July 1940 Driver Evans, RASC, became the first fatality when his parachute failed to function correctly.[65] Parachuting was an inherently dangerous pastime even under perfect conditions. Under military conditions, in an experimental environment and under time pressure, fatalities were inevitable. There were forty-six parachute training deaths at Ringway between Driver Evans's demise and 8 March 1945. This represents a minute percentage of the estimated 60,000 men (maybe equating to approximately 300,000 jumps) that were trained during the corresponding period. However, the proportion of fatalities to jumps was much higher than this during the initial stages. There were five deaths in the first 6,532 jumps up until 20 June 1941.[66] The reaction to fatalities was mixed. Not surprisingly Gale considered fatal accidents sickening, dreaded by all but somewhat inevitable.[67] Frost was more philosophical, 'It always seemed that some curious or freakish fault had occurred when a man fell to his death. I think most of us felt that this was inevitable at the beginning and that only by trial and error could perfection be attained; in any case these things only happened to other people and we ourselves were perfectly safe.'[68] Before the war the War Office had considered parachuting 'a precarious occupation' the danger of which might render it undesirable during peacetime.[69] By 1940 the Air Ministry believed that the accident rate was a justified risk when considered against the much higher level of risk paratroops would be exposed to once they had landed and were in the midst of battle.[70]

However, no matter what the official attitude was, at No. 1 PTS training fatalities were having a discernable effect. After the first three deaths Captain Lindsay on the staff at Ringway recorded that 'morale in the [No. 2] commando was not good with distrust of the parachutes.'[71] Although fatalities had a strong morale effect they did not significantly affect the training rate, owing to the small numbers. Nonetheless, for every death there were many accidents causing injuries of varying degrees which took many men away from training for weeks, months or even permanently. Jumping through the floor apertures of bomber aircraft brought its own dangers, as pupils were liable to hit their heads and faces on the way out. Sometimes this was superficial, with the medics at Ringway being called on to stitch approximately five facial injuries a week. Paratroops who left the aircraft semi-conscious or unconscious due to 'ringing the bell' were liable to break ankles and legs on landing. Even a good jump could result in broken limbs if the landing was on uneven ground.[72] On top of injuries there was also a high proportion of refusals to jump amongst the 'tough' men of No. 2 Commando. Of the first 342 men to be trained 30 refused to jump. Although not considered a disgrace by any means, a refusal to jump did result in a man being removed from the parachute course. Refusals were considered indicative of low morale and were apt to be contagious in a unit and therefore every effort had to be made to keep them to a minimum.[73]

A new approach to training was required to reduce fatalities, the general accident rate and the number of men who were succumbing to their natural psychological instincts by refusing to leave the aircraft. Another factor was also driving the need for alternative training methods. There was an acute shortage of aircraft available at No. 1 PTS during the war for all those reasons described in the previous chapter. A lack of aircraft was recognised as a constraining factor on the training rate almost as soon as Ringway opened for business.⁷⁴ The six Whitleys initially allocated to the PTS were barely adequate to cope with the requirements of training for No. 2 Commando. As the training requirement expanded the complement of aircraft available for training did not increase at a similar rate. In March 1942, when the additional battalions for 1 Parachute Brigade were being trained the establishment was increased to twelve Whitleys and an Avro Anson for liaison duties.⁷⁵ The Air Ministry was always reluctant to release aircraft to Ringway for training. Following Operations TORCH and HUSKY the requirement to train from the C-47 Dakota became obvious. The War Office made a request to this effect but the Air Ministry regarded it as inopportune at that time to release any Dakotas to the PTS although it promised to bear the appeal in mind in the future.⁷⁶ As the Operational Record Book of the CLE records, No. 1 PTS 'conceived in haste and born in a spirit of mental confusion, always lacking sufficient staff, accommodation and equipment, soon learnt the art of making shift.'⁷⁷

That art was practiced to great effect in No. 1 PTS and manifested itself as the ingenious application of synthetic training. Synthetic training of any description was still a new concept at the beginning of the war. Applying synthetic methods to any training process required a deep understanding of the system of instruction and the required end state. With parachute training still in its infancy a high degree of vision and confidence was necessary. Nevertheless some of the methods, devices and apparatus developed at Ringway during the early part of the war were so successful that many are still used today in one form or another by airborne forces around the world. That is not to say that failures were totally non-existent. One of the first methods instigated was to make trainees jump from the back of a moving truck on to a grass verge to simulate a parachute landing. As one instructor at the PTS recalls, 'I think the grass was just as hard as the tarmac. It didn't encourage us very much and we had lots of accidents and it was stopped.'⁷⁸

Static training apparatus, housed in the hangars at Ringway, proved highly successful and popular. In order to attain a training output rate of 100 paratroops per week the CLE listed the required equipment as eight mock-up apertures, eight trapezes, four low jumping towers, four aircraft wind engines and eight trestled fuselages equipped with panels, strops and lights.⁷⁹ All of this equipment was designed and built by the Technical Development Section (TDS) and workshops at Ringway. The trapeze resembled a suspended aerial maypole from which trainees could swing and then, on letting go free-fall a short distance and roll as they hit the ground. Platforms were constructed approximately eight feet high on the sidewalls of the hangars which students pushed them selves from, also to practise their roll when hitting the ground. Slides were constructed, propelling trainees from nine feet up and releasing them at speed about three feet from the ground, again to practise the necessary landing drills.

A high platform, approximately twenty feet up, was fitted with a parachute harness attached to a cable wound round a drum fitted with paddles. The student jumped from the platform and the paddles slowed his descent as the cable unwound, approximating the speed of a real parachute drop. There were also mock apertures built to resemble the hole in the floor of a Whitley through which a trainee could practise his exit. All of these pieces of equipment (some of which were given enticing names such as 'the gallows') allowed soldiers to practise the techniques required on landing without the risk of significant injury. It also allowed the entire process of parachuting, from leaving the aircraft to landing, to be broken down into a series of drills that could be learned by rote.[80]

Mock-up fuselages were accurately constructed to represent the interior of various aircraft. In these students could practise the procedures for hooking up their parachutes and the drills to be followed immediately prior to leaving the aircraft. This released actual aircraft for live parachuting that might otherwise have been kept on the ground to cater for these periods of instruction. All of this apparatus could be used simultaneously by different groups of students at various points of their training. Much of the synthetic equipment was housed in hangars, which allowed it to be used despite the vagaries of the weather. Another innovation that released aircraft flying hours was the use of a tethered balloon from which parachute drops could be conducted. The idea of parachuting from a stationary platform as opposed to a moving aircraft was not completely unique. The Russians and the Poles had built towers all across their countries pre-war in order to allow civilians to experience parachuting. The balloon was a more flexible solution to the tower as it was not permanently fixed in one spot.[81] The advantage of jumping from the balloon was that it allowed the trainee to exit the aperture in the floor of the balloon cage without being buffeted by the slipstream that would be present with an aircraft. This reduced the risk of tumbling and of serious if not fatal accidents occurring.[82] Also a balloon drop allowed an instructor on the ground to criticise a student's technique via a megaphone as he descended. There were disadvantages; the lack of noise and bustle associated with an aircraft fuselage made the silent balloon cage a far more contemplative environment immediately before jumping. As the commanding officer of No. 1 PTS commented, 'In the case of the balloon everything seemed so much more cold-blooded, and on first experience many men regarded the procedure as positively diabolical.'[83]

As well as relieving aircraft flying hours this intensive and scientific ground training, followed by two jumps from the balloon, reduced the accident rate, if the number of fatalities can be considered indicative. Following the first five fatalities from 6,532 jumps the next death did not occur until more than a further 5,000 jumps had been conducted, in September 1941. The next fatal accident after this did not take place until plus of a further 13,000 jumps had occurred, in January 1942. The accident rate levelled off during 1943 and despite the much higher training rate only nine deaths occurred throughout 1944 and 1945.[84] A by-product of synthetic training was a reduction of the psychological impact of the physical act of parachuting. This was largely due to the procedures being reduced to a drill, which on the appropriate command became an involuntary reaction. MacDonald Hastings, a feature journalist with 'Picture Post', recalls this experience, 'It was his voice, not my muscles that did it. I went in spite of myself.'[85] Another trainee elaborated on the effect when exiting from the balloon, 'I was jumping No. 2 and sitting on the edge of that aperture I was convinced that I would never be able to launch myself into space. I looked up at the instructor's hand as he said to me "good exit No. 2" and I still did not believe that I could get out of that cage, but all the aperture training took over and on the word "GO" my body reacted automatically and out I went.'[86]

In order to maintain this psychological advantage paratroops had to continue to jump at regular intervals after their basic training. It was recognised that if men did not jump frequently then they could lose their nerve.[87] Also new techniques were developed throughout the war and new equipment, such as leg bags and valises were introduced which required further instruction. Therefore a programme of continuation training was required. By the time the second and third parachute battalions were being established in late 1941 the ideal basic training course for *ab initio* paratroops was considered to be fourteen days, during which time two balloon jumps and five descents from aircraft were completed. Courses could be shortened *in extremis* to deal with sudden influxes of soldiers requiring training. Each of these standard courses took 240 trainees split into 4 syndicates of 60 each. This therefore resulted in approximately 200 trained paratroops leaving Ringway every two weeks or an average of 100 per week.[88] This figure had been identified as the output capacity of No. 1 PTS from its inception although it represented a significant target against the original training rate, which saw only 500 members of 11 SAS trained in approximately eight months, or an approximate average of less than 20 per week.[89]

It was considered that there was potential for growth in the training capacity at Ringway and that the 100 per week output standard set in late 1941 could be quadrupled in order to achieve 5,000 trained paratroops by May 1942, the equivalent of approximately two-and-a-half parachute brigades. However this remained a projected training rate only.[90] In fact approximately 2,500 men were trained by mid-1942, representing a low average actual training rate between March 1941 and July 1942 of less than 50 paratroops per week. However, it was sufficient to supply the formations that had been established at that time and the training rate did increase significantly during this period so that at the end of 1941 it had reached the projected 100 per week.[91] Also in mid-1942 the feasibility of establishing a second airborne division was under consideration. To achieve this by the end of 1942 a further 5,500 paratroops would have to be trained in six months, equating to a training rate of just over 200 per week.[92] This would have been achievable by increasing the establishment of No. 1 PTS and allowing courses to overlap and run concurrently. There was also capacity in parachute schools that were being established in other theatres such as the Indian Parachute School at Chaklala near Rawalpindi and No. 4 Middle East Training School, originally at Kabrit in the Canal Zone of Egypt. Much of 4 Parachute Brigade was trained in India and the Middle East. In the event a second airborne division was not established until spring 1943 and so no drastic increase in the training rate was necessary and courses at No. 1 PTS continued at 200 to 250 trainees at any one time (approximately 100 per week on average) through to early 1944.[93]

By June 1944 Britain's airborne forces in Europe were fully established at two divisions and an independent brigade. From then on training was only required for battle casualty replacements and for continuation training and, as a result, the output rate eased somewhat.[94] During this time an important principle was recognised which linked training to the readiness of formations. For example, two allied airborne task forces, each of approximately 1,500 paratroops were being held in the Mediterranean in preparation for the invasion of the south of France. One of these task forces was held at one week's notice to move, the other at three weeks' notice to move. This allowed training resources to be prioritised and directed at the formation on the shortest notice to move in recognition of the fact that the second formation would have a further two weeks of dedicated training once the first was deployed.[95] This method was also used at Ringway and priorities were continually revised and allocated in order to ensure that training was delivered where it was most urgently required. The priority shifted back and forth between 1st and 6th Airborne Division between January 1944 and mid-1945, dependent upon which division was assigned to operations next.

The individual training of paratroops was not a major factor in the progress of airborne development during the war. The initial training rate was slow but was sufficient to meet the requirements of the equally steady pace of recruitment. The lack of aircraft released to the training establishment did initially have an effect on the throughput at the CLS and CLE. However, when rapid expansion began, the training apparatus was sufficient to meet the increased requirement. Critical factors were the ingenuity required and willingness to adopt and develop novel training methods such as synthetic training. This maximised the use of resources, in particular aircraft and time, and ensured the efficient processing of manpower by reducing the injury and fatality rate and overcoming the psychological barriers that caused men to refuse to jump. Therefore a high proportion of the men recruited for parachuting duties became trained paratroops. The pass rate for initial parachute courses at No. 1 PTS appears to have been between 85 and 95 per cent throughout the war.[96] The training regime was developed while the required output rate was low so that when the requirement was raised the PTS was able to respond effectively. Priorities were allocated and switched between formations allowing efficient targeting of training to cope with the particular conditions at different stages of the war.

Glider Pilot Selection and Training
Besides paratroops the other specialist human resource that would be vital to the development of British airborne forces was glider pilots. The selection and training of glider pilots presented very different challenges to that of the parachute troops. The numbers involved were much lower but the technical aspects of training were far more demanding.[97] There were two main obstacles that had a detrimental effect on selection and training. The glider pilot, more than any other area of airborne forces, straddled the spheres of influence of both the Army and the RAF. Because of this their selection and role bore the brunt of inter-service rivalry between the War Office and the Air Ministry. This complicated and slowed the initial selection process. Also the training regime for glider pilots was far less robust than that of paratroops. Because of this outside influences could very quickly have an effect on the training output rate.

The Glider Pilot Regiment (GPR) was unique among British Army units. Conceived and established during the war it was obsolete, declined and disbanded shortly after victory. The qualities required of glider pilots, certainly according to the Army, were unique. Brigadier George Chatterton, the commander of the glider pilots for most of the war summed up the requirement, 'It is the most unusual unit ever conceived by the British Army. A soldier who will pilot an aircraft, and then fight in the battle, a task indeed.'[98] The initial recruits, like those for parachuting, were drawn from No. 2 Commando, therefore the standards required were again those of the commandos and not specific to this new discipline. The only added requirement was that they should have some flying experience.[99] Chatterton set out the qualities he believed were required, including high general standards, intelligence, initiative, the ability to command and competence in all methods of warfare.[100] In late 1940 the CLE formally listed the requirements for a successful glider pilot in order of importance. These were judged to be initiative, fearlessness, intelligence, robustness, flying experience with powered aircraft and flying experience with glider aircraft.[101] Already, soldierly qualities were being ranked above those of airmanship. The CLE had determined that it was better to select a proficient soldier and train him as a pilot than *vice versa*. This proved in time to be the more efficient approach to the requirement. Also at this time procedures for processing volunteer glider pilots were published. This involved Army commands calling for volunteers, followed by an interview of the applicants by the Senior Air Staff Officer (SASO) of the respective command. Recruits were sent to selected RAF squadrons for a medical examination and finally the names of successful volunteers were to be forwarded to HQ No. 22 Group of AC Comd for transmission to the War Office.[102] These procedures and standards were not revised until the period of expansion in late 1941. It was not until 26 September 1941 that glider pilot selection was ordered to conform to the mental and physical standards of RAF aircrews. A joint RAF/Army interview board was also convened.[103]

The psychological factors affecting glider pilots were different to those of paratroops. It was unlikely that a glider pilot would find himself initially isolated on the battlefield, nor did he have to overcome the reactions against jumping from an aircraft. However, the effect of mental dislocation between an airfield in England and a battlefield on the Continent was still apparent and, if anything, exacerbated by the task of the glider pilot. Once again Chatterton sums up this concern.

> I assumed that flying a glider on tow must impose a severe strain on the pilot, a strain that would in no way be alleviated by the knowledge that on landing in the battlefield there would be no prospect of an immediate return. Now, I knew that on landing, after flying a powered aircraft for a long time, there is an overwhelming desire to relax completely and to sleep. Such facilities are available for pilots landing at an air station; but, I asked myself, what would happen to the pilot who landed in the middle of battle?[104]

The glider pilot would have to cope not only with a change in environment but also a severe change of role. Being recruited as a proficient soldier would help him to do so. He would find it more natural to revert to his initial role than an airman would to adapt to the relatively unnatural tasks of an infantry soldier. Potential glider pilots were induced to volunteer by many of the same incentives as paratroopers, although it appears there was not quite the same level of direct propaganda. Officers who qualified would be paid an extra two shillings per day. Other ranks would be paid the same on initial qualification but this would be on a sliding scale, dependent on rank and experience up to an extra three shillings and sixpence a day.[105] Glider pilots wore the distinctive maroon beret and with the formal establishment of the GPR in February 1942 they had their own cap badge. The question of a unique flying badge for glider pilots was raised in late 1940 but the issue was clouded by spurious inter-service considerations.[106] It was not until a year later that the Air Ministry agreed that a winged badge could be worn on the left breast akin to the RAF,

Personnel and Training 111

although it did object to the 'bilious red' of the prototype badge submitted for approval.[107] As well as these material incentives the War Office also ran a carefully targeted direct recruitment campaign. Soldiers who had previously passed aircrew selection with the RAF but for whatever reason had ended up in the Army were identified, traced and given a direct opportunity to volunteer for the GPR. Most men approached in this manner appear to have been only too keen to accept.[108]

Due to these schemes and incentives and because of the low numbers involved, there were no undue problems producing the manpower required to fulfil the glider pilot role. However, the invidiousness of basic inter-service friction was to have a stalling effect on the initial production of glider pilots and take up an unnecessary amount of staff effort. The General Staff firmly believed that a glider pilot had to be a soldier capable of being effective in battle once he had landed his troops and equipment. They should be capable of grouping together on the landing zone into formed units in order to carry out specific military tasks such as conducting special raid parties, special weapon parties or acting in the general light infantry role.[109] They should be capable of joining in the battle and even leading the troops that they had just landed. A joint committee had very quickly agreed with this view and confirmed that glider pilots should be Army soldiers trained by the RAF.[110] Understandably though, the Air Staff emphasised the difficulty of landing gliders *en masse* at a pre-determined place and time, at night, on unknown ground and possibly in the face of enemy fire. It logically concluded that, 'This is no task for a beginner in flying who is primarily a soldier.'[111] On 11 December 1940 Air Marshal Sir Arthur Harris made the following statement that was to become infamous within the GPR.

> The idea that semi-skilled, unpicked personnel (infantry corporals have, I believe, even been suggested) could with a maximum of training be entrusted with the piloting of these troop carriers is fantastic. Their operation is the equivalent to forced landing the largest sized aircraft without engine aid – than which there is no higher test of piloting skill.[112]

As they proved themselves on operations the glider pilots wore this quote as a badge of honour.[113]

The General Staff refuted the Air Staff assessment of the difficulty of glider flying. They advised that the gliders in development would have an elaborate arrangement of flaps and air brakes so that control of the aircraft and the approach to the landing point would be a relatively straight forward operation. The reasons for the pilots being experienced soldiers capable of command were reiterated.

> The glider coxswain [pilot] on touching down will be the only man present who will know exactly where the landing has been made and

in which direction the troops should go. He has the best forward view, he is highly trained in map reading and studying ground from the air, and he will have noted the lie of the land to the objective. Even if only a Corporal, he will be the one to lead the other 23 officers and men to the right place.[114]

Once again the Air Ministry would not concede that using soldiers as pilots was a satisfactory arrangement. However, the Air Staff was in danger of arguing itself into a corner. The RAF did not have pilots to spare at the end of 1940. It was already arguing against the development of airborne forces due to the drain it would have on the bomber effort. It could not then in good faith now offer bomber pilots to fly gliders. The Air Staff came up with increasingly desperate solutions to get around this. It was suggested that 'war-weary and ex-operational pilots' who could be quickly converted to glider flying might be suitable. The psychological impact on such aircrew of deliberately landing in the middle of a battle does not appear to have been considered.[115]

One of the arguments against the General Staff's insistence on combat-trained pilots could have been that it was not a universally held position. When America began to train glider pilots they did not recognise the requirement for them also to be fighting soldiers.[116] This approach did have some advantages. A glider pilot who had not been trained to fight would be unlikely to commit himself to combat and on landing would be more inclined to preserve his own safety by making for a designated RV. This scheme would help conserve an extremely valuable human resource. However the General Staff did not adopt the American scheme, which was later proved to be detrimental on operations. During Operation MARKET GARDEN over 1,000 American glider pilots who could have been usefully employed in the battle sat ineffectively around Eindhoven and Nijmegen waiting for evacuation.[117]

Despite this the Air Staff maintained its position that glider pilots must come from the RAF into 1941. A scheme was proposed where operational bomber pilots would be taken off front line duties very briefly and given the minimum training required to enable them to fly gliders. They would then return to their bomber squadrons and form a virtual pool of glider pilots ready for airborne operations as required.[118] However, this approach overlooked the need for combined training, for glider pilots to regularly train *en masse* and with airlanding infantry and supporting arms. This proposition also ignored the probable requirement for maximum bomber effort immediately prior to a major operation that might employ an airborne element. It also contradicted the Air Staff's own assessment of the difficulty attached to flying and landing gliders. As expansion of the airborne capability began in mid-1941 and more experience was gained in operating gliders and the tactics associated with airlanding units, the logic of the General Staff's argument became inescapable. In August 1941 the Air Ministry finally acceded and agreed that glider pilots should be fighting soldiers. They further agreed that they could be officers or NCOs and that they would be seconded to the RAF for training.[119]

This decision had taken over a year to reach and consumed considerable staff effort. During this time the recruitment of glider pilots had been slow as neither service wanted to commit time and effort to a scheme that might then have to be handed over to the other. The direction and process of training also suffered. It was difficult to formulate a firm training policy when the basis of ownership was in dispute. The Air Staff, keen to protect aircraft from secondary activities, continued to argue whether specialist training was required at all. Why establish an elaborate and costly glider training system when existing RAF pilots could be quickly converted immediately prior to a planned airborne operation?[120]

Despite this protracted dispute an initial glider section was established at Ringway as early as 23 August 1940.[121] Three pilots were brought from the Special Duties Flight at Christchurch, which had been established to conduct radar experiments on gliders for anti-invasion purposes. Squadron Leader Robert Fender and two colleagues landed at Ringway unannounced in two Avro 504s on 8 August 1940. A few days later four single-seat gliders arrived by road and were re-assembled.[122] Once at Ringway the glider section's remit continued to be experimentation; it was not designed to conduct training. It was accepted that potential glider pilots would first have to receive tuition on powered aircraft in order to learn basic airmanship principles. The RAF's Elementary Flying Training Schools (EFTS) were full to capacity and therefore the initial intake of glider pilots was attached to Army Cooperation squadrons for their initial training. This began in November 1940. The syllabus was limited to achieving the ability to complete at least three successful solo landings without using the engine.[123] The first twelve pilots were ready to progress to glider training by February 1941 having been graded from average through competent to very steady.[124] As the glider training establishment grew through the war this initial powered flight phase became more formal. Space was made in EFTSs around the country, such as those at Burnaston (No. 16), Booker (No. 21) and Shellingford (No. 3), and an official syllabus for a twelve-week course with eighty hours of flying was drafted.[125] Later in the war glider pilots were drawn from EFTSs as far away as Canada and the US, such as No. 1 Basic Flying Training School (BFTS) at Terrell, Texas and No. 3 BFTS in Miami, Oklahoma.[126] This phase of training was essential to ground new pilots in airmanship and to allow initial training in an aircraft that, with power available, was inherently more forgiving than a glider. An aircraft with power also allowed more take offs and landings to be practised than a glider might manage in the same time. However, despite this necessity, eighty hours flying over a twelve-week period could be considered to represent an excessively comprehensive programme.

The first dedicated Glider Training School (GTS) had been opened at Haddenham, subsequently renamed Thame in December 1940. It was established by moving limited resources from Ringway. This included just a handful of converted single-seat civilian gliders and five aging Tiger Moths for tugs. Fifteen members of the RAF with pre-war gliding experience had been drafted in as instructors.[127] On 12 March 1941 the first pupils arrived at No. 1 GTS by which time a single Hotspur was available. The initial syllabus allowed six weeks for thirty-one hours flying. This was later cut to four weeks and approximately fifteen hours flying time. As the system expanded No. 2 GTS was opened at Weston-on-the-Green in December 1941 followed subsequently by No. 3 at Stoke Orchard, No. 4 at Kidlington and No. 5 at Shobden.[128]

The GTS only qualified the pilot to fly the Hotspur, a non-operational glider. Conversion and further training were required on the Horsa and took place at Glider Operational Training Units (GOTU). Nos. 101 and 102 GOTUs were established on 1 January and 1 February 1942 at Netheravon and Kidlington respectively. Pilots spent a further four weeks at a GOTU learning to operate the troop-carrying glider. The advent of the Hamilcar brought the requirement for further conversion and to cope with this Heavy Glider Conversion Units (HGCU) were opened from mid-1942 at Shrewton and Fairford and later at Peplow and North Luffenham. HGCU's required pilots to train for a further two weeks.[129] This meant that the entire glider training process took eighteen weeks, not taking into account delays, movement between sites and administration.

However, the length of the training system had been anticipated to a certain extent and was not a problem in itself. The greatest challenge arose from the complexity of the system. There were up to four separate stages (EFTS, GTS, GOTU and HGCU), which were of different lengths, making it difficult to coordinate the output from one course with the intake of another. Also the courses at different stages were designed to cope with different numbers of pupils at any given time. Add to this the geographic dispersion of many of the training units and the result was a training process that was unduly complex which in turn led to fragility. This weakness meant that any outside pressure or influence could cause the system to stall and begin to break down.

The training process proved to be very disjointed. For instance in early 1942 a single EFTS (No. 16, Derby) had an output of 45 pilots every three weeks following their twelve-week course. The two GTSs available at that time could only take 16 pupils at each school once per fortnight for a six-week course. The output of one phase of the system did not match the intake requirements of the next part. Hence gaps or bottlenecks were likely to occur.[130] Changes in policy, often reactive rather than considered, exacerbated this effect. In mid-1941 the War Office requirement for glider pilots stood at 800 but was reduced to 600 later that year.[131] This reduction in numbers led to a decrease or pause in recruiting. This in turn meant that there was at one stage in mid-1942 temporarily insufficient potential pilots at the glider pilot depot in Tilshead to feed the input for the EFTSs. This meant either accepting gaps that would have a knock on effect through the system and leave valuable instructors and aircraft standing idle for up to a fortnight or reducing the output of the EFTSs by a significant amount to keep the system ticking over. The Air Ministry was understandably unhappy with this situation considering the resources invested in the glider pilot flying system.[132]

Bottlenecks could also quickly occur. The most obvious cause was the number of suitable gliders and tugs available for training. In early 1942 faulty manufacture of parts slowed the production of Hotspurs. This impeded the training rate to such an extent that raising the matter to ministerial level was considered. At the same time this problem was compounded by difficulties in securing tug aircraft. Glider training was more expensive in terms of flying hours than parachute training. A single parachute training sortie could benefit between ten (with the Whitley) and twenty-five (with the Dakota) students. A single glider training sortie could only ever benefit a maximum of two students. Aircraft shortages impacted on the GTSs, which then had subsequent impact on the GOTUs.[133] The GOTUs had been established before the Horsa had been cleared for flying with troops on board and had to initially make do with the Hotspur. The initial production of Horsas lagged behind schedule and in mid-1942 this was having an effect on the output of the GOTUs.[134] As numbers going through the GOTUs slowed so those waiting to move forward from the GTSs and EFTSs were held up in the system. This had impact all the way back to the GPR depot in Tilshead where potential pilots were held inactive and prone to the influence of Army 'bull'. The fragility of the system led to these gaps and bottlenecks persisting as the process attempted to compensate. This prompted one group of glider pilots to make a complaint over the hiatus in training and the excess of spit and polish to their local Member of Parliament, Mr D.N. Pritt. Pritt in turn brought the grievances to the attention of Churchill. The Prime Minister was unimpressed by Pritt's representations and considered him to be 'a bad man' for raising the matter. [135] Nevertheless the Minister for Air, the Minister for War, CIGS and CAS were asked to comment. Sinclair admitted that there were problems with the training process and blamed them on the uncertainty in overall airborne policy that had existed in late 1942. He maintained that the situation would improve as the coordination of airborne training became more robust.[136]

Despite the inherent vulnerability of the training system there is no evidence that there was ever a significant shortfall in glider pilots available for operations. This was due to the relatively low numbers required and the incremental nature of the increase in the airlanding organisation. In September 1940 the number of glider pilots required was initially estimated as 350. The Air Ministry was confident that these could be trained by the spring of 1941.[137] This proved to be over optimistic, and by April 1941, of the sixty-six soldiers originally identified for training, only thirty-seven had demonstrated the necessary aptitude, of whom twenty-four were undergoing glider training.[138] Even this initial low estimate of the requirement, which was proving difficult to achieve, had been formulated against flawed assumptions. First it was assumed that the wastage rate in glider pilots during operations would be low. This was proved to be a false hope, although the casualty rate did decrease during the course of the war.[139] The second supposition was that each glider would require one pilot.[140] This was not an unreasonable assumption as American airborne forces flew gliders with a single pilot throughout the war.[141] However, the slow start to British glider training allowed more time for experimentation and through this it became clear that two pilots were required. The second pilot had a number of duties, including map reading to the landing point, calling out altitude and distance during the approach, making radio and intercom calls, liaison with the passengers during flight, taking some of the physical strain from the first pilot during long tows and taking control if the first pilot was injured or killed.[142]

The correction of these false assumptions resulted in a recalculation and a sudden increase in the requirement for glider pilots. In May 1942 the total was set at 640, to be reached by 31 December 1942.[143] Just two months later the Air Ministry was working towards a target of 1,200 glider pilots trained by 1 July 1943.[144] The expansion of the airborne establishment through this period necessitated a further increase in the training rate. However, the decision to commit to the recruiting and training of further pilots was delayed by the COS Committee in order to assess the results of Operation HUSKY. Following the operation, judged as a success despite the losses, the Air Ministry agreed to train a further 800 men by June 1944.[145] By D-Day the training rate reduced and levelled off but in July 1944 the question of RAF personnel being converted to the glider role was raised again. The overall casualty rate for Operation OVERLORD had been lower than expected. Prior to the invasion the Air Staff had offered to transfer approximately 10,000 men from the RAF deferred list to the Army. The War Office, however, was content that it could produce the monthly intake of 135 trainees required to maintain the glider pilot pool from its own resources.[146] The Air Ministry pointed out that this situation was irrational. The RAF had thousands of surplus and potential aircrew on the deferred list and yet the Army was putting inordinate effort into recruiting 135 soldiers suitable for flying every month. These would then require many weeks' training before they would be ready for operations. Additional to this the RAF had 1,500 RAF personnel, many of them pilot instructors, tied up in the glider pilot training system. In an act of commendable pragmatism the Air Ministry was offering a more efficient solution to the production of glider pilots. The Air Staff even displayed sensitivity towards the concerns of the Glider Pilot Regiment and was keen not to dilute their prestige by a sudden influx of RAF pilots.[147]

The General Staff accepted the Air Ministry's offer and on 28 August 1944 ACAS (Plans) ordered the conversion of 228 surplus RAF pilots to the glider role. This was to take place in October and November 1944.[148] The decision was timely as the scale of casualties from Operation MARKET and the future requirements for airborne forces in Southeast Asia Command (SEAC) were becoming clear. Approximately 830 glider pilots did not return from Arnhem, the majority becoming prisoners of war. Of the 500 who were available for further operations the balance was uneven between first and second pilots, resulting in only 230 complete crews being available. The War Office estimated an immediate shortfall of at least 335 pilots for future operations in all theatres.[149] In order to maintain the glider pilot establishment it was stated that a monthly output of 160 trained pilots would be required, to be achieved by training eighty Army and eighty RAF personnel. Detailed plans were drawn up to maintain this rate through to February 1945.[150] By that time there were 708 RAF glider crews held in readiness within SEAC.[151] However, it had become clear that the rapid advance of allied ground forces across Burma had resulted in planned glider operations becoming unnecessary. The original plan for Operation DRACULA, the assault on Rangoon, was revised and the glider element was discarded. This signalled the end of the requirement for British glider pilots in the theatre. Members of the RAF and the GPR returned to Britain as the war in Europe ended.[152]

Operation VARSITY was the final airborne operation of the war in Europe. The casualty rate amongst glider pilots was relatively light, and had further operations been mounted, such as Operation JUBILANT, a plan to secure German POW camps to prevent atrocities being committed, there were sufficient glider pilots to sustain the requirement.

The recruiting and training of glider pilots did not affect the rate of development of British airborne forces during the Second World War. This was despite the fragility of the training system that was employed. Success was due in large part to the relatively low numbers required throughout the war. As with paratroops, the low level of operational activity from mid-1940 to mid-1943 allowed the training system to be developed and refined and for a reserve of pilots to be built up before they were required for Operation HUSKY and then OVERLORD. The War Office remained open to suggestions to improve the efficiency of the training system. When it became clear that the RAF could provide glider pilots at less cost than recruiting them from the Army, the General Staff accepted the idea, despite having fought so hard to ensure that pilots would be soldiers only a few years before. It has been commonly assumed that the War Office was forced to accept RAF pilots due to the high casualty rate sustained at Arnhem.[153] However, it has been shown that this policy was put in place before MARKET GARDEN and was a pragmatic solution to the efficient employment of resources. This pragmatism helped ensure an adequate supply of pilots throughout the war.

The build up to the invasion of Sicily provides the only consistent example of poor training directly affecting glider operations. It did not affect the number of pilots available but rather the quality of their performance. This was due to a number of factors. The training rate in England during the winter of 1942 had been low, due in part to a lack of tug aircraft. Some glider pilots were reduced to just seven flying hours throughout the entire winter period.[154] Further to this, the difficulty in moving Horsa gliders overseas meant that most of the British glider pilots would have to fly the Hadrian into battle, an aircraft that they were not familiar with. This was exacerbated by an extremely short period available for training in North Africa prior to the invasion of Sicily. Glider pilots arrived by sea in Algeria late in April 1943, leaving only two months for training in theatre prior to the invasion. This period was further truncated once the gliders had been assembled and administrative moves across North Africa had been carried out. During the time available a typical training programme was just fourteen short flights of only ten minutes each.[155] Chatterton believed the lack of training had a detrimental effect on the morale of his pilots during the build-up to the invasion.[156]

Finally and most critically, the lack of RAF tug aircraft in North Africa meant that the glider pilots who would fly 1 Airlanding Brigade on to Sicily had to be towed by aircraft and pilots from the USAAF Troop Carrier Command. There was very little time for the British glider pilots to conduct collective training with their American tug pilots. This was the primary factor in the poor military effectiveness of British gliders during HUSKY, an operation that emphasised the importance of collective training.

The Coordination of Collective Training and Exercises
The training examined thus far concentrated on the individual, allowing him and his comrades to be delivered to the battlefield efficiently. This phase of training did not in itself prepare airborne soldiers for the battle once they landed. Commanders often stressed this point, highlighting parachuting and gliding as only a means to an end. 'Every paratroop has instilled into him the fact that jumping from an aircraft is merely an incident on the way to battle.... It is not allowed to assume greater importance in his mental make up. Every operational paratrooper, no matter what position in a unit he may fill, is first and foremost a fighting soldier and is trained as such.'[157] Some commentators extrapolated this

view and postulated that paratroops and airlanding infantry would require no form of specialist training beyond their delivery, as once on the ground they would fight as conventional infantry.[158] This was not a sound argument, a fact identified early by the CLE. It recognised the unique tactical circumstances that airborne troops would find themselves in post-landing. Probably dispersed, they would have to find their way in unfamiliar country, possibly to a distant objective. They would have to fight with open flanks and an unguarded rear. Due to the lightweight nature of their arms they would have to generate speed during the attack in order to produce sufficient momentum of force. For these reasons it was considered that airborne units and formations would require specialist tactics and training.[159] Training at unit and formation level would require close coordination across all arms and services to produce combat effectiveness. This principle had gained increased credence between the wars and been succinctly expounded by CIGS General Sir George Milne in 1927. 'It is the co-operation of all necessary arms that wins battles and that is your basis for training in the future. I want that to be your principle in training – combination and co-operation of arms.'[160]

Airlanding infantry would need to coordinate their training with that of the glider pilots. Infantry, including airlanding and paratroops, would need to train with their supporting arms, artillery, engineers, reconnaissance and logistics, to build a high level of mutual understanding.[161] Airborne forces would need to train alongside the conventional units and formations with which close liaison would be required during operations. Above all these Army units and formations would require close coordination with the RAF to ensure their efficient delivery on to the battlefield. Initially, like so many other aspects of Britain's airborne forces, this coordination function was entrusted to the CLE. However, the location at Ringway generated doubt as to whether the CLE would be able to deliver the breadth and intensity of training required. Its position on the Cheshire Plain made it particularly vulnerable to the bad weather often prevalent in that area. Numerous days of rain made drying parachutes and other drop equipment difficult. The above average mean wind speeds often curtailed parachute and glider training. Nevertheless, time and resources were being heavily invested in Ringway during this period, making a sudden change of location difficult. The CLE already occupied 68,700 square feet of covered floor space by November 1940. The relatively free and unthreatened airspace around Manchester, and the extensive runways and accommodation, more than made up for the vagaries of the weather and Ringway remained the centre of coordination, certainly for parachute training, throughout the war.[162] The CLE, under the administrative supervision of HQ No. 22 Group RAF quickly established its roles and delineated internal responsibilities. Training was divided into three areas: individual parachute and glider flying training, individual battle training and collective training and exercises. It was initially envisioned that the CLE would coordinate and control all of these activities.[163]

The staff at Ringway was directly tasked with organising cooperational training across all branches of airborne forces and with conventional units and formations. This prompted calls for the 'considerable expansion' of the training organisation in order to keep pace with the enlargement of airborne forces through 1941.[164] This expansion was often rapid, and, as a result haphazard, leading to tension in the wider training establishment. The glider exercise organisation, created to allow coordinated training of glider pilots and airlanding troops, suffered under dual control. The training itself was controlled through the CLE initially, but the tow aircraft required to facilitate that training still came under Bomber Command. The situation became more complicated once control of glider exercise units was devolved down to the airborne divisions.[165] This situation also extended to collective parachute training. The requirement for dedicated RAF assets to support airborne training became apparent. This principle had been adopted on a small scale in mid-1941 at the CLE with the formation of the first glider exercise flight. The flight was established with up to ten Hotspur gliders and up to ten light aircraft (Hectors or Westland Lysanders) to act as dedicated tugs.[166]

The level of training of the bomber crews that took part in Operation COLOSSUS in February 1941 was criticised by the CLE. Some of the pilots had only four days to absorb the intricacies of parachute dropping by night in hostile terrain and weather, and although the operational drop was relatively successful, substantial errors were apparent during the final rehearsal.[167] This and the general expansion through 1941 caused permanent RAF squadrons to be formed to solely service the operational parachute and airlanding brigades.[168] On 1 January 1942 296 Squadron RAF was formed as a dedicated glider support squadron. The initial establishment was drawn up as sixteen Wellingtons with fifteen complete aircrews. Assuming 75 per cent availability this would have meant that twelve Horsas could be towed at any time, carrying around 300 troops or approximately two companies in any one lift. The aircrew for 296 Squadron were to be regarded as a select body of glider experts who would be able to assist in training, experimentation and the development of airlanding tactics.[169] 297 Squadron was also established to service operational parachute training, and, later in January 1942, 38 Wing RAF was created as a headquarters to the RAF airborne training establishment, able to coordinate the requirements of the developing 1st Airborne Division HQ.[170]

38 Wing was the manifestation of an idea that had been formulated in the Air Ministry, championed by Browning and enthusiastically accepted at the highest levels, including Churchill. This was the concept of a 'Nucleus Force', a group of RAF aircrews who would be in constant training with airborne forces and would also be available with their aircraft

for short-notice, small-scale operations. A lengthy paper written in HQ 38 Wing RAF convincingly argued the efficiency of expanding that organisation so that operations as well as training could be managed without having to divert resources from Bomber Command. The aircraft would also be available for air re-supply operations and *in extremis* could bolster the bombing effort. Browning took this paper to CIGS, CAS and the Prime Minister who, in Browning's words, 'clutched at it as a drowning man would a straw, and demanded that more copies of the paper be prepared.'[171]

In July 1942 295 and 298 Squadrons were ordered to form as additional units for 38 Wing. It was assessed that these four squadrons in total would be sufficient to meet contemporary airborne training requirements, although the formation of a second division might have required further expansion.[172] In October 1943 38 Wing was expanded and became 38 Group RAF. The Group contained nine squadrons: four equipped with the Albermarle, four with the Stirling and one with the Halifax.

In January 1944 a second RAF Group, No. 46, was formed. No. 46 Group belonged to Transport Command and was tasked to conduct normal transport duties, but when the demands of airborne operations and training arose would come under command No. 38 Group. It had an establishment of 150 Dakotas. These two groups remained the principal air support formations to Britain's airborne forces through to the end of the war.[173] During this period of expansion the Air Ministry also produced detailed air tactical doctrine to support airborne operations. This work emphasised the importance of joint training and rehearsals with airborne forces, particularly during the period leading up to operations.[174] Specific training tasks and standards were stated for aircrews taking part in airborne operations.[175] The aircrews of 38 and 46 Groups took to their new role with 'a total commitment' and 'felt properly trained for all airborne duties, although these did require flying skills not normally demanded of other aircrews'.[176]

The development of the RAF component dedicated to airborne forces was initially slow. This is unsurprising considering the Air Ministry's resistance to the progress of the concept and the shortage of suitable and available aircraft. It took the first eighteen months of development up to January 1942 to establish just two RAF squadrons allocated to parachute and glider training. Following the consideration of the 38 Wing paper by the Prime Minister the rate of expansion increased and was given added momentum by the decisions made at Casablanca and the experiences in North Africa and Sicily. Hence in the twenty-two months following January 1942 a further seven squadrons were added to the airborne establishment and the services of the second Group, No. 46 were also made available. Therefore by October 1943 the dedicated aircraft available were sufficient to service the onset of training for the invasion of Normandy.

Despite these advances and obvious enthusiasm, the combined training of all arms of the airborne division was still not straightforward. The operational integration of airlanding formations with parachute formations provoked prolonged doctrinal debate, which will be examined later. However, it was quickly grasped as early as September 1940 that airlanding troops and paratroops would have to train together if they were to operate together to any degree. Their roles were envisioned as distinct but mutually supportive and therefore required close coordination.[177] Training together on a routine basis was not as simple as it first appeared, owing to the geographic dispersion of parachute and airlanding units across Britain. A subset of this problem was the difficulty in integrating airlanding infantry and supporting arms with their glider transports during training. Gliding was inherently dangerous, even during training. A poor landing with a mixed load of heavy equipment and troops could have fatal consequences. During the training for D-Day there were seven fatal accidents to glider pilots, three where carried troops were killed also. Chatterton witnessed one such event.

> The last glider to come in, a Hamilcar, came in somewhat fast, at about 110 miles an hour. She bounced, half took off again, landed again and bounced, then careered across the airfield and crashed into a group of Nissen huts, which disintegrated. The two pilots in their cockpit remained on top of the rubble, but the tank, which must have hit the building at eighty miles an hour, went straight on before coming to a stop some fifty yards away.[178]

Such a level of risk to equipment and, more importantly, to trained personnel was considered unacceptable during training. Remarkably this situation could lead to airlanding troops gaining their first experience of landing with full equipment during their first operation.[179] This was clearly not an ideal situation and in attempted mitigation glider pilots flew with representative loads (often blocks of concrete) while airlanding troops practised unloading equipment from static, pre-positioned gliders. However, this could not recreate the adrenalin of a landing or the difficulty of unloading from a glider that may have partially broken up on landing.

Despite these challenges airborne training could be successfully coordinated for specific purposes and events, even if it was difficult to integrate all the facets on a routine basis. This success is observable through the numerous airborne forces collective training events and exercises that were successfully conducted during the course of the war. These exercises essentially took three different forms. First there were low level exercises involving only a section to a company of airborne troops, designed either as demonstrations for conventional troops, senior officers and politicians or to introduce a realistic enemy into larger anti-invasion exercises. Second there were exercises from company to divisional level, designed to assess and practise the generic airborne tactical principles and techniques that could be required during any operation. Finally there were exercises from platoon to battalion level that were conducted as focused rehearsals for very specific airborne operations.

Exercises in the first category began very early in the airborne development process. One of the first took place on 26 October 1940 near Macclesfield. Two gliders landed in a small field next to their supposed objective, a railway viaduct. The exercise was observed by Lieutenant Colonel Stephenson of SD4 in the War Office.[180] By mid-April 1941 11 SAS Battalion had taken part in six exercises. These included one conducted at Tatton Park for the benefit of the CIGS, General Sir John Dill, on 12 December 1940 incorporating a drop by sixteen paratroops with a landing by five single-seat gliders.

In January 1941 the CLE sent a detachment to the Aldershot area to take part in a Home Forces exercise named DRAGON. According to the orders issued by Ringway the airborne portion of the exercise was designed to demonstrate 'how troops may be transported by air and surprise pin-point landings made at pre-determined places by parachutes and gliders'. The demonstration was small but configured to show the C-in-C Home Forces and other senior officers what a larger force might achieve. Five single-seat gliders and one Whitley carrying a section of paratroops were used to represent a landing by seventy-five airlanding troops and two platoons of paratroops respectively. This time the objective, in a premonition of Operation FRESHMAN, was a small enemy factory producing a new kind of weapon. During April 1941 several sections from 11 SAS Battalion represented German parachute forces during a Northern Command exercise named ALFRED. They were dropped in the path of 1 Corps to delay and harass their movement north as they advanced to attack a German beachhead in the area of Middlesborough.[181]

These examples are typical of the early exercise activity conducted by Britain's fledgling airborne forces. The commander of No. 1 PTS, Group Captain Maurice Newnham, believed they were valuable in providing a realistic training opportunity for the troops involved.[182] There is some merit in this view; however these small-scale exercises did little to progress tactical development. This lack of focused collective training led to false procedures and tactics being developed. An example is the 'Landing Drill for Sub-Sections' produced in November 1940. The five-page document listed in great detail the roles and equipment to be carried by each member of a parachute section. Further to this it prescribed precise actions for each section member on landing. Which man was responsible for unloading each container, who carried which ammunition loads and who was responsible for protection were all listed. Even the author had to admit that the procedures represented 'a long rigmerole [sic]'.[183] This approach demonstrated a fundamental ignorance of the nature of airborne warfare at the tactical level. The drills developed to ensure successful demonstrations in front of important visitors were given tacit tactical validity. This ignored the fact that probable confusion on a combat DZ did not lend itself to being resolved by a series of drills. The increased use of tactical drills had gained acceptance through 1941 in order to meet the needs of training thousands of conscripts and wartime volunteers. This method recognised the dilution of the pre-war professional Army and was aimed to suit the lowest common denominator.[184] Airborne forces, through selection and rigorous individual training, were soldiers of higher ability than average. Flexibility and initiative were the requirements to ensure effective action post-landing and these qualities required repeated and varied collective tactical training and exercises in order to develop.

By mid-1942 collective training was being planned by the airborne HQs rather than being coordinated by the CLE. However it was only being conducted at company level. Training in any greater numbers required special arrangements and requests had to be judged individually on their merits.[185] In May 1942 Browning produced a paper requesting the resources to allow a battalion of paratroops and a battalion of airlanding troops to exercise simultaneously.[186] Shortly afterwards 295 and 298 squadrons RAF were added to 38 Wing to allow this to happen. It was subsequently decided that the minimum requirement for collective training should be three live exercises involving aircraft per man every two months. It was accepted that many of these exercises would be small scale but others would have to be on a large scale and involve cooperation with conventional troops in order to achieve the training objectives. Taking into account the effects of the weather it was determined that this training rate would result in approximately twenty exercises per month being conducted for the airborne establishment. From these calculations a precise aircraft requirement could be deduced which confirmed the establishment of 38 Wing RAF.[187]

Despite this more scientific approach to the requirements of collective training, the practicalities of integrating all arms was still problematic. Conducting exercises for different arms and units made it difficult to balance training and ensure it was constructive for all involved. The danger of conducting glider exercises with live loads has already been explained. Incorporating the artillery into exercises was also difficult. Live firing gunnery exercises had to be separated from troops on exercise because of the obvious danger. Due to the technical complexity of achieving accurate gunnery, static, live-firing target practice often took precedence over integration with larger airborne exercises. Captain Wilkinson RA does not recall taking part in any exercises involving infantry participants. His training consisted mainly of firing practice on various ranges across the country.[188]

This difficulty in integrating all arms was not unique to airborne forces, and to overcome the problems, planning and procedures were often practised by conducting TEWTs (Tactical Exercise Without Troops). These were essentially tabletop map-based exercises that trained the planning staff without actually requiring troops deploying into the field. As early as 1938 theoretical groupings of parachute troops had been drawn up to be used in anti-invasion TEWTs.[189] In 1941 a United States military observer, Lieutenant Colonel J.C. Kennedy, reported on a TEWT representing an airborne assault on the Humberside coast involving 500 gliders being released fifteen miles off shore to land on three separate LZs. The force delivered included infantry and artillery.[190] Prior to D-Day Crookenden recalled a TEWT being conducted for over 100 officers from 6 Airlanding Brigade.[191] However, while this type of training was of benefit to formation and unit staff there was still no substitute for live exercises to practise and develop tactics throughout the airborne establishment.

Larger scale tactical exercises helped to develop and practise specific operational techniques and brought out important lessons that could then be applied during battle. One such operational technique that required coordinated exercises to provide effective training was the glider pilot's 'funnel'. This system taught pilots to fly into an area 1,000 feet above the ground and 1,000 yards out from the LZ perimeter. From this position a pilot had sufficient height, speed and manoeuvrability to land at any point on the LZ up to 2,000 yards inside the perimeter. This training taught the pilots a scientific method of applying flaps during a landing and allowed the pilot to select the most tactically appropriate landing point. Practising this tactic as a single glider was relatively simple. It required dozens of gliders to be released simultaneously to provide a realistic test of the pilots' skill. These principles were adhered to during training and as a result the glider pilots improved in tactical ability and confidence.[192]

In some cases an exercise could highlight a fault in tactics or procedures. An exercise held by 1 Parachute Brigade near Grimsby a few weeks before Operation MARKET GARDEN demonstrated the limitations of the issued signals equipment in built-up areas. This lesson allowed the signals organisation to reassess its equipment holdings and subsequently fly into Arnhem with a greater than usual supply of field telephones and cable.[193] Despite these important results some exercise participants were often unimpressed and were more likely to remember the exercise names, such as THRUST, BIZZ, MUSH and BANGER than they were the tactical objective of the activity.[194]

Nevertheless, these exercises were all vital in developing and practicing general airborne tactics. However once units and formations were warned for operations much more specific training was required in order to prepare them for the tasks they were likely to execute. As early as October 1943, in preparation for D-Day Gale designated a number of likely missions

6th Airborne Division would have to undertake during the invasion. These were divided into offensive and defensive tasks. The attack missions included the capture of enemy gun batteries and strong points, the capture of tactical ground behind the beaches and defiles and river crossings, and the attack of enemy reserves. Defensive tasks included holding tactical features, bridges, defiles and high ground overlooking the beachhead. Gale's directive also listed lessons that had been identified during previous exercises and split the tasks down into explicit training objectives. For example, the assault on an enemy gun battery required the following training objectives to be practised.

(i) The discovery by patrols of whether the pos[itio]n is occupied.
(ii) Practice in finding the way to an objective after landing in the dark.
(iii) The re-org[anisation] after the assault and the rapid assembly at the pre-arranged R.V.[195]

This represents a relatively sophisticated approach to training. Complicated missions are reduced to a number of simpler tasks each of which are then given a related training objective. Each of these training objectives could therefore be achieved in isolation as time allowed and still contribute towards the success of the overall mission.

As an operation became imminent and objectives were revealed, planning could be conducted down to unit and sub-unit level. Those units and sub-units could then focus their training very precisely on achieving their own specific objectives. In conventional land-bound operations leaders at all levels could have an opportunity to see the ground they were due to fight over. Prior to offensive operations the lie of the land could be studied from observation points or during reconnaissance activity. During defensive operations the ground might be occupied already and would be intimately familiar to all the defenders.

Airborne leaders and soldiers did not have this luxury. The ground they would fight over might be hundreds of miles away over completely unfamiliar terrain with foreign features. Very few would be able to fly over the ground prior to an operation; the risks were too high. Most leaders would have to make do with maps and aerial photographs. The information available in these formats could be enhanced by combining the two and building scale models. The use of models was not unique to airborne operations but they were used extensively prior to all levels of action from minor raids to divisional assaults.[196] In one case this technique was ingeniously adapted to provide a highly sophisticated training aid.

A model of the Orne bridgehead was built to assist the glider pilots prior to D-Day. A No. 38 Group officer, Squadron Leader Lawrence Wright conceived the idea of making a film of the model. Chatterton recalled the results. 'By using a blue filter on his 16 mm Camera, and by holding the camera at a calculated height, he produced a remarkable film for briefing. It gave a complete picture of a glider coming in from 1,000 feet, down to about 100 feet, and provided a most realistic impression of what the landing in Normandy might be like.'[197] The camera was rigged from the ceiling using a series of wires. Chatterton maintains that the film was of great assistance to the pilots. Staff Sergeant Jim Wallwork, the commander of the first glider to land in Normandy, recalls seeing the film just once or twice and considers that it was of novelty value only.[198] While this initiative represented a very modern approach, along with all models it really only represented an aid to briefing as opposed to practical training.

Training in an environment designed to imitate a known objective is an activity militarily defined as rehearsals. Rehearsals were recognised as an important aspect of training early in the airborne development process.[199] For rehearsals to be valuable they had to be conducted in an environment that replicated the real objective as closely as possible. Due to this, rehearsals were only really applicable to small-scale actions, up to battalion level. Above this size realistic environments were more difficult to create and their effectiveness was therefore reduced. This meant that rehearsals were normally restricted to training for raids or for discrete actions planned as part of a wider operation.

Rehearsals can be broken down into two broad types. First they could be conducted over ground closely resembling that over which the operation would be executed. Alternatively full-scale models could be built to replicate features that would be encountered during an operation.

In some cases the ground selected for a rehearsal might have resembled the area of an operation only in the very broadest sense. An example of this was the final rehearsal for Operation BITING that took place on the cliffs around Lulworth on the Dorset coast.[200] In other instances great care was taken to select a piece of ground that replicated the target as closely as possible. Pine-Coffin recalled during the build-up to Operation OVERLORD training over two water obstacles at Countess Weare near Exeter that were fairly similar to the Caen Canal and the Orne River in Normandy.[201] Chatterton claimed the credit for finding another piece of real estate 'near Hinton, Buckland and Bampton Aston which almost completely corresponded to the area of the River Orne and the Caen Canal'.[202] Exercise CANDID was run over this part of the Oxfordshire countryside in May 1944 during which members of the GPR were put through a 'complete rehearsal' of their operational role post-landing on D-Day. This included finding their way to a designated RV and then moving along a specified route across the two bridges to their evacuation point on the beach near Ouistreham. Map traces show that the area of the exercise almost exactly replicated the ground to be covered during the operation, including the distances between RVs, the relative positions of the bridges, the routes in between and some landmarks such as woods.[203]

Where representative ground could not be found or where a very specific location was the objective of an operation then full scale replicas were sometimes built. The sophistication of the construction required varied. Operation DEADSTICK was the first act of D-Day, requiring six Horsa gliders to land in very close proximity to the River Orne and Caen Canal bridges so they could be seized by *coup de main* by a reinforced company of the Oxfordshire and Buckinghamshire Light Infantry. Two LZs were selected for this operation, 'X' and 'Y', between the river and the canal. Both were very constricted by their size and surroundings. Seven glider crews had been selected for the operation. Mock-ups of LZ 'X' and 'Y' were marked out at Netheravon on Salisbury Plain using white tape. They exactly replicated the LZs in Normandy in size, shape and relationship to each other. During rehearsals the gliders were released at the correct point with reference to the landing points. The gliders then flew the correct flight paths to land in the taped areas. This was first conducted in daylight, then with the pilots wearing dark goggles, then at night using flares and then using just moonlight as they would have to during the operation.[204] The link between training and the performance of glider pilots had been made clear during the preparation for Operation HUSKY. The training for operation DEADSTICK was intensive. The six crews and one reserve crew allocated to the operation flew forty landings in the three months before D-Day. The 300 crews taking part in Operation MALLARD, the mass glider landing on the evening of D-Day had on average just ten day and two night landings in the six months prior to D-Day.[205] Wallwork is modest about Operation DEADSTICK's success and is, 'at pains to emphasise that any member of the Regiment could have done it, given the practice and attention we lucky ones experienced.'[206]

Some mock-ups required considerably more effort than marking an area on the ground with tape. Britain's first airborne operation used a mock-up. A full-scale wooden model of the Tregino aqueduct was built in Tatton Park close to Ringway.[207] Major General Tony Deane-Drummond, a Lieutenant on Operation COLOSSUS, 'regarded the mock-ups as essential aids but mainly for briefing purposes' as opposed to being used for live rehearsals.[208]

Alongside Operation DEADSTICK, 9 Battalion, The Parachute Regiment under command of Otway attacked a German coastal battery at Merville during the early hours of D-Day as part of Operation TONGA. This was a complex battalion operation. A complete mock-up was built below Inkpen Ridge in Berkshire and the four casemates were constructed out of wood and canvas. The surrounding wire, ditches and minefields were all replicated. Otway's battalion rehearsed repeatedly on the mock-up for months. Many different permutations of attack were practised until any man could fulfil any other's role and nearly all eventualities had been covered. However this was not the only result of these rehearsals.

Churchill inspects early paratroopers giving an unimpressive display at Ringway on 26 April 1941.

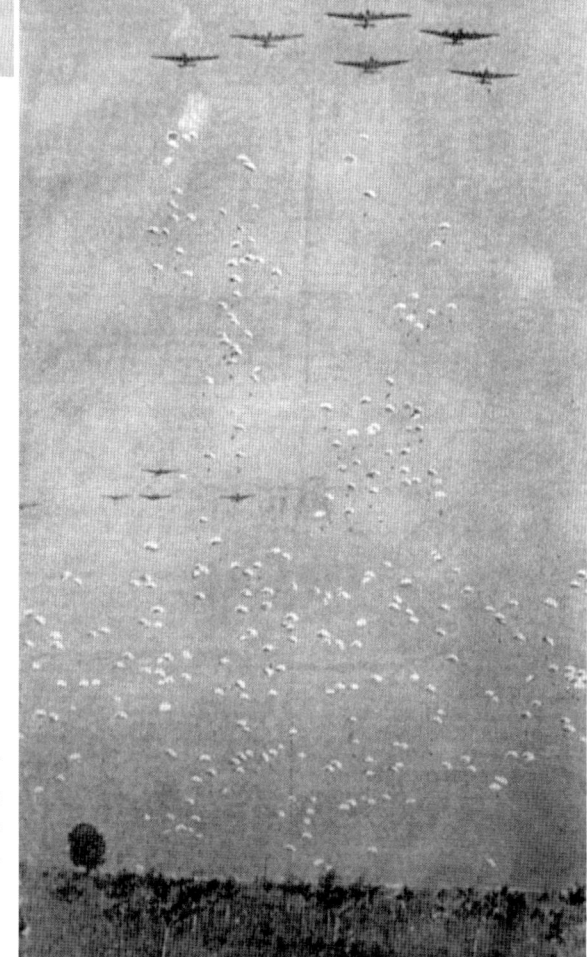

'A most spectacular performance ...' The *Daily Telegraph* reports on Soviet airborne manoeuvres near Kiev on 26 October 1935.

Major General Browning shares a joke with Churchill during the Prime Minister's second visit to Ringway on 16 April 1942.

Airborne gunners firing the 3.7-inch mountain howitzer during training in October 1942 before the introduction of the American 75mm pack howitzer.

Lieutenant General Browning (with gloves, cane and beret) observes equipment trials. Note the American airborne officer on the far left of the photograph.

The X-type parachute. Sixty-eight square yards of silk, a valuable commodity throughout the war.

'There is, I'm afraid, no alternative...' The Armstrong-Whitworth Whitley Mark V.

'Calculated to damp the light hearted enthusiasm with which these young men volunteered...' Paratroopers inside the cramped fuselage of the Whitley.

The Harrow, Hertfordshire and Bombay: all unsuccessful candidates for service with British airborne forces.

Whitley and Dakota conduct a simultaneous drop for comparison.

Airspeed Horsa ready for assembly, clearly demonstrating why it was impractical to transport glider by road.

The glider that never was. The innovative Slingsby Hengist did not make it into full production due to the success of the Horsa.

The General Aircraft Hamilcar heavy glider undertaking loading trials with a Universal Carrier.

Positive propaganda. An early portrayal of a paratrooper for general public interest.

BRITISH PARATROOP READY FOR ACTION

...ain is training some of the best of her troops in the specialist arts and duties of the ...chutist. When dropping from a transport plane an overall is worn over his uniform. ...on the ground this is discarded and the paratroop, as above, is ready for action

Paratroopers undergoing basic tactical training on the moors near Hardwick Hall.

'No thinking, but plenty of action.' Hardwick Hall tested the physical and mental endurance of paratroopers, instilling confidence and independence as a result.

Trainee paratroopers undergo synthetic training on one of the highly successful rigs developed at Ringway.

'I went in spite of myself.' Jumping becomes an involuntary action as a paratrooper exits the side door of a Dakota complete with leg bag.

'Cold blooded and positively diabolical.' Preparing to jump from the balloon at Tatton Park.

A paratrooper makes a training descent from the balloon.

The Indian airborne project, established in 'an atmosphere of lethargy and indifference'. Indian paratroopers emplaned for a training jump at Chaklala.

...rainee glider pilot receives individual training in the General Aircraft Hotspur.

...early glider pilot course poses for the camera at Kidlington. A rare photograph of Lieutenant ...onel Rock, killed in a gliding accident in October 1942.

Glider pilots undergoing tactical instruction prior to collective training.

'A soldier who will pilot an aircraft, and then fight in the battle, a task indeed.' Glider pilots form up ready to fight as an infantry unit.

The value of collective training is manifest in the obvious familiarity between these glider pilots and their tug crews.

Paratroops fit parachutes before emplaning in Dakotas for a large scale exercise in April 1944, in preparation for D-Day.

'The imperative of example.' Major General Eric Bols, commander of 6 Airborne Division receives instruction from Squadron Leader Kilkenny, chief instructor at No. 1 Parachute Training School.

Airborne pioneers from two services. From left to right, Brigadier Gale, Squadron Leader Harvey, Wing Commander Newnham, Lieutenant Colonel Down and Lieutenant Colonel Hope-Thompson

'The distillation of the airborne establishment.' Lieutenant General Frederick 'Boy' Browning.

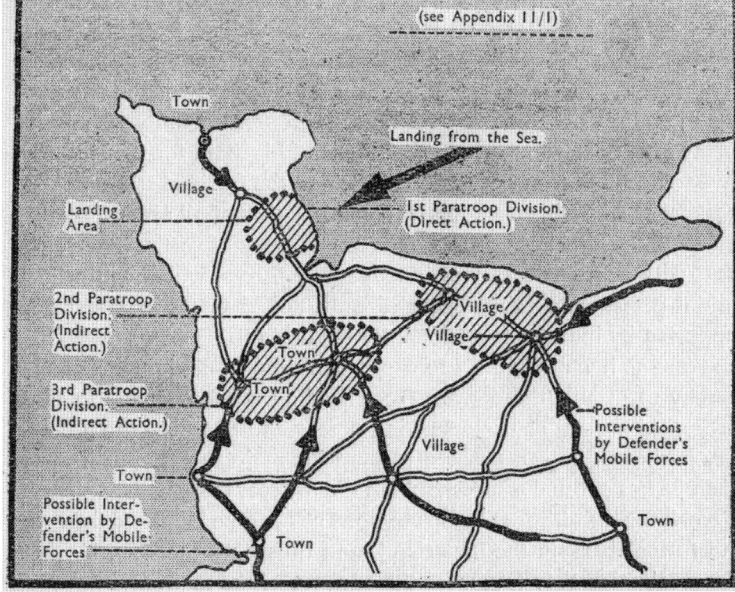

Miksche's 1943 attempt to explain airborne doctrine. This diagram's uncanny resemblance to the plans for D-Day did not go unnoticed by the authorities.

Tactical misuse. A mortar team from 2 Independent Parachute Brigade employed as conventional infantry in Italy in April 1944.

A contemporary cartoon candidly depicting the multiple influences on airborne development, including the Ministry of Aircraft Production (MAP), Airborne Forces Experimental Establishment (AFEE) and Army Co-operation Command (ACC).

ASSISTED TAKE-OFF

Lieutenant Alan Jefferson recalled an additional psychological effect. 'The whole Battalion was keyed up to fever pitch: the barometer of morale was absenteeism, and that registered zero. Each man had been made to feel indispensable and consequently he felt important.'[209] The conduct of rehearsals also helped to alleviate the psychological pressure of dislocation for paratroopers and glider pilots on landing. A piece of ground or a feature that was familiar would have been a point of focus on landing, allowing the individual to instinctively proceed on his mission despite being in foreign and hostile territory. The ground, objective and routine became so familiar that when the operation itself was finally executed it often seemed to the participants like yet another iteration of the training and rehearsal process. Crookenden described his experience of this effect on landing on D-Day, 'Everything was exactly as I had imagined it would be, as, thanks to our [training] I never had to think more than a moment where I was', and again during his flight on the way to land on the east bank of the Rhine, 'I think that we all of us half expect to see the familiar roofs of our barracks below us when we jump out.'[210] Familiarity through rehearsal helped mitigate the effects of dislocation.

The coordination of training was critical to convert individually trained personnel into a militarily effective fighting force. Combining the training of the various arms of the Army element of the airborne force presented particular technical challenges. Some of these were common to conventional formation training: geographical dispersion of units and the problems of integrating artillery into an all arms scenario. Others were unique to airborne forces, such as the risks in exercising glider landings with live loads. However bringing the crucial RAF support to the training process involved overcoming institutional and systemic barriers. In actual fact this appears to have been overcome in a most pragmatic manner. Inter-service rivalry, apparent between the respective ministries and in the higher commands was bypassed and ignored at squadron, group, battalion and brigade level. This willingness to cooperate meant that effective training programmes could be formulated including conducting tactical exercises at all levels. The evidence suggests that these airborne exercises were planned and conducted in a serious manner. This can be explained by two facts. First is the high value of flying hours that the exercise planners could not be seen to be wasting. Second is the very real risk of death or serious injury to those exercise participants being deployed by parachute or glider. The mitigation of this risk required careful planning by the staff and meticulous preparation by the participants. This focused application permeated the airborne establishment and resulted in a zealous approach to the conduct of exercises. These were essential in demonstrating the fledgling airborne capability and in developing tactical doctrine and rehearsing specific operations.

The Human Resource as a factor in Development

Effective manning was one of the keys to success of the development of Britain's airborne capability. The recruitment and training of airborne troops progressed remarkably efficiently in spite of the many hurdles and challenges encountered during the process. It has been suggested that the manpower invested in various special formations during the Second World War could only barely be justified by their operational impact. However the figures quoted for the airborne establishment are not significant in terms of overall Army manning. Nevertheless, due to operational casualties the constant recruitment and training requirement was a formidable task. Added to this quantitative challenge there were also qualitative considerations that inflated the effect of the drain of manpower into the airborne capability. Immediately post D-Day there were 16,623 personnel on the strength of the Army Air Corps (AAC), which represents less than three-quarters of a per cent of the British Army's total strength at that time. However the strength of the AAC is not necessarily an accurate measure of the size of Britain's airborne forces. The figure includes SAS troops, which were not strictly airborne forces in the truest sense, and simultaneously omits all those arms and services, such as Royal Engineers and Royal Artillery, who formed a vital component of the airborne capability.

At its height the total strength of the airborne force, i.e. two divisions and one independent brigade, has been calculated at 28,345. The addition of 44th Indian Airborne Division would bring this number up to around 40,000. Even this total represents only just over one-and-a-half per cent of the Army's total.[211] However this figure does not paint the full picture. In common with many other elite organisations, Britain's airborne forces relied on the infantry to generate most of its fighting power. At maximum strength the airborne capability contained approximately 14,000 infantrymen, or around twenty standard infantry battalions, a resource bill that the infantry found increasingly difficult to pay, particularly after June 1944. The value of the human resources absorbed represent a greater investment than these small percentages indicates due to the generally higher calibre of the men it employed. There is some sympathy for commanders such as Montgomery and Slim who, along with Horrocks, contested the validity of skimming off the best soldiers to fill specialist requirements such as airborne forces.[212]

These figures only represent a snap-shot in time and therefore do not tell the full story of the need for ongoing recruitment. This challenge was exacerbated by the tendency to retain airborne forces in combat, using them as regular infantry once an airborne operation was complete, a task unsuitable for such lightly equipped formations. This occurred during Operation TORCH where 1 Parachute Brigade, having taken part in parachute operations in November 1942, were retained to fight as conventional infantry until April 1943, losing 1,700 casualties during the five month period. In Italy 2 Independent Parachute Brigade was employed as a conventional formation for most of the campaign, although relatively light casualties resulted. In Normandy 6th Airborne Division remained committed to defensive operations until August 1944 causing 4,457 casualties in three months. Even airborne operations that were relatively limited in terms of duration could generate significant casualties. Operation HUSKY caused 706 casualties within 1st Airborne Division over four days and the same formation sustained another 7,167 casualties during Operation MARKET GARDEN. 6th Airborne Division took 1,434 casualties over three days during Operation VARSITY in March 1945 and then marched to the Baltic over the next two months.[213] Overall Britain's airborne establishment based in Europe lost approximately 16,000 casualties over a two-and-a-half year period. This equates to well over half the total strength of the airborne force at its peak.

Apart from the physical effect of the casualty rate there was also a potential moral effect. This could be manifested as an increase in battle fatigue and a corresponding reduction in morale within a unit or formation. However, any reference to the typical indicators of a break down in morale, such as insubordination, absenteeism and psychological breakdown, are completely absent from both the primary and secondary sources.[214] Even during the desperate defence of Oosterbeek, as casualties mounted and there was no chance of victory, the evidence suggests that morale was maintained, as demonstrated by the defiant manner of those airborne soldiers that became prisoners of war and by the discipline maintained during the final withdrawal. Maintenance of morale in the face of the high casualty rate is perhaps explained by the appreciation by airborne troops that they were an elite force. The training, the beret and the badges fed this attitude to the point where airborne troops almost gloried in adversity. This corresponds with Montgomery's belief that self-respect and regimental tradition were basic and contributory factors to high morale.[215] This can be seen in the rhetoric that can often be found in the published primary sources.

> The airborne have blazed their trail of glory and sacrifice across the deep vault of heavens, and over the blasted soil of Europe. Spawned from this war, their memory will live on forever and new men will come forward to take over and carry on the unfurled flag with the sign of the Pegasus upon it, fluttering proudly in the breeze of freedom, ever ready to spread their wings over Europe again, or even further if the call should come "Defend thy home by attack." They did not die in vain.[216]

Notwithstanding the maintenance of morale, physical losses still had to be replaced to ensure the combat effectiveness of the airborne capability. That this was achieved with relatively few problems was mainly through the pragmatism of the War Office and the Air Ministry. Although firm principles were laid down for the recruitment of paratroops (volunteers only) and glider pilots (soldiers only) these were willingly amended when it became clear there was a more efficient or effective solution.

Pragmatism often implies a level of compromise. Although hand selected volunteers may have provided the best material for paratroops in ideal circumstances the exigency of rapid expansion in 1942 and 1943 necessitated compromise, and the wholesale conversion of battalions with back-filling by volunteers was accepted. Early tactical experimentation proved that trained soldiers made the most effective glider pilots; however a compromise was agreed and RAF pilots were accepted in 1944 when it became clear that this represented a more efficient method of generating the required manpower. There is no evidence that either of these compromises adversely affected operational effectiveness.

While pragmatism was the key to recruiting, the conduct of individual training required a different approach. There was no model available to the initial proponents of airborne warfare. Ideally the process of training had to be envisioned before it was implemented. In reality this had to happen simultaneously and a high degree of ingenuity was required by the trainers. Many innovations were implemented and applied to the individual training of paratroops. These included synthetic training methods, many of which were ahead of their time and hence have persisted to the current day. This ingenuity had many beneficial consequences, including the mitigation of the effect of aircraft shortages on training and of some of the psychological pressures of airborne warfare. However the application of ingenuity was not so apparent in the training of glider pilots, perhaps because the rate of glider production and doctrinal ambiguity meant that the glider component continually lagged behind the parachute component, which led to staff continually struggling to keep up.

Airborne collective training was approached in a focused and pragmatic matter. A high degree of coordination was achieved across all arms and, more importantly, across the two services, which translated into close and instinctive coordination on the battlefield in most circumstances. Where a lack of coordination between the Army and RAF appears to have impinged on operational effectiveness it can often be explained by a lack of collective training preceding the event. This was certainly the case in North Africa and Sicily, although at Arnhem it was due to other factors, which will be examined in the following chapter.

It has been suggested that, in general across the Army, there was an over-reliance on artillery preparation during exercises to the detriment of low-level infantry tactics and the problem of 'the last 200 yards'. Further to this it has been argued that collective training in conventional infantry units and formations minimised the need for low-level initiative and decision-making and excluded any preparation for deviation from a set drill or plan.[217] Due to their unique tactical situation this was certainly not the case with airborne collective training and exercises. A lack of fire support and the light armament of an airborne unit meant that it had to carry an objective through speed and the closest contact with the enemy. 'The last 200 yards' could not be ignored or left to the artillery and it had to be practised and rehearsed to ensure success. Although the use of mock-ups and models was not unique they did have added significance.

One criticism of British Army training during the war is that it concentrated far too much on inducing immediate and exact obedience to orders and very little else. This was a reflection of the Army's basic doctrine with an emphasis on centralised control.[218] Low-level initiative and decision making was essential to ensure that an airborne soldier could first rendezvous with his comrades and then carry an objective no matter what configuration of section or sub-unit he found himself in. The repeated attacks on the Tregino and Merville mock-ups and the bridges at Countess Weare were designed to give the soldier the experience and confidence to tackle his objective no matter what the circumstances. Although perhaps true of the conventional infantry, the assertion that 'major exercises were not occasions on which minor tactics were practiced' is not applicable to the same degree for airborne training, where emphasis was always placed on the individual and his primary group.

The provision of personnel and training could easily have proved a stumbling block on the path of airborne development. Aircraft shortages and political friction had the potential to hinder or even halt the recruiting and training process. Nevertheless the human resource for the airborne capability steadily increased at probably the maximum rate that could be expected. However, despite the successful provision of trained manpower it does have to be recognised that the process was incremental. By June 1941 there were just 338 paratroops, or half a battalion available for operations.[219] In October 1941 31 Independent Brigade Group was re-roled as 1 Airlanding Brigade with a strength of approximately 3,000, and in March 1942 with 1 Parachute Brigade operational the total available airborne manpower had risen to around 5,500 personnel. 1st Airborne Division reached full strength in December 1942 and with 2 Independent Parachute Brigade available the figure reached 15,500. Finally, in February 1944, with the operational readiness of 6th Airborne Division the full total of 28,000 airborne troops available in Europe was reached.[220] Thus, despite the success of airborne recruiting and training, the size of an airborne force that could be utilised at any particular stage of the war was constrained by the manpower available at the time, a significant factor in the development of the airborne concept of operations, as will be seen.

CHAPTER FIVE

Command and Control

Leadership in Airborne Warfare

As J.F.C. Fuller observed, men are more than 'mobile tripods to rest rifles on'.[1] The value of an individual soldier on the battlefield lies in the fact that he is a sensate being and can act according to his own will and judgement. However, during intense conflict where death or injury might be judged as imminent these factors can become a disadvantage. Training and discipline will help overcome a soldier's natural instinct towards self-preservation but they are seldom enough in themselves to force a soldier to act against his will. Soldiers, and more importantly groups of soldiers, need to be effectively led and motivated in order to achieve their assigned objectives in a cohesive manner. Leadership and motivation are elements of the less tangible moral component of fighting power. Strong motivation is the result of effective leadership and effective leadership is a function of a good commander.[2]

Some of the characteristics of Britain's new and evolving airborne capability made conventional approaches to command and control incongruous at best and detrimental to effectiveness in some circumstances. Command from platoon level upwards in the British Army was routinely invested in officers. Those who served with airborne forces, apart from the specific processes described in the previous chapter, underwent the same selection and training as all other officers.[3] At first glance it may appear that an airborne officer, and in particular a parachute officer, would have a more difficult task than his conventional infantry or armoured counterpart. Before he could lead his men in battle and motivate them to fight he had to lead them out of an aircraft in flight, urging them to risk their lives in the process. In actual fact airborne training and deployment into battle gave an airborne commander an advantage in leadership. Airborne training, hazardous in itself, gave an airborne leader a shared experience with his men, a point of reference and a demonstration that he was willing to accept the same risks as the soldiers he led. A resolute determination to share the discomforts and dangers of his subordinates will always make leadership and motivation a simpler task.[4] A parachute platoon commander was normally the first man out of the door of a Dakota during training and operations, an evident example of leading from the front to those following behind him. An officer in an airlanding unit had to sit face-to-face with his men in the back of a Horsa as they all risked death or injury as they glided into battle. Senior officers could not exempt themselves from this and brigade and divisional commanders arrived on the battlefield in a glider or below a parachute in the same manner as a private soldier. These are very obvious manifestations of John Keegan's 'imperative of example' and produced a deep bond between an airborne officer and his men.[5] Despite the advantages of an unavoidable kinship of danger, airborne leaders also had another difficulty to contend with. Airborne soldiers were deliberately inculcated with an ethos of personal initiative. In order that the advantages of this approach were not stifled leaders would have to develop a style of command that would allow their soldiers a degree of autonomy.

At more senior levels of command the 'imperative of example' became less easy to demonstrate and less important to those being commanded. Soldiers needed to be convinced that their senior generals knew their business if they were to have any confidence in his abilities. Professional knowledge or understanding nearly always appears in any one of the many lists of principles of leadership.[6] However, a senior commander during the Second World War was unlikely to have any direct experience of working with airborne forces. Certainly during the first half of the war a senior commander's knowledge of airborne warfare was likely to be purely theoretical, based on his own study, research and understanding. Some senior officers were more liable to apply themselves to this pursuit of knowledge than others. However there was a further, more vital factor that senior commanders would have to adapt to in order to successfully command an airborne force. Airborne warfare relied entirely on the contribution of the air force. Senior commanders and their staff would have to work very closely with air force counterparts and ideally joint HQs were required to effectively coordinate the two services.

Commanders, like their soldiers, are individuals and will therefore each adopt their own approach to command. Thus there is an element of unpredictability as to how any commander thinks or acts in any given situation. The character of the commander at any level can therefore have a direct influence on the interpretation of doctrine, the application of tactics and ultimately the outcome of a battle. As Gale commented, command is 'one of the most evasive subjects – certainly one of the most important', not only because of the immediate effect of a commander's decisions on the current battle but because of the influence that he can exert well beyond the scope and tenure of his command.[7]

The Experience and Influence of Superior Commanders

Until November 1942 Britain's airborne forces were employed on raiding operations. These were planned, mounted and directed by DCO and Combined Operations, which had built up some experience on the subject since 1940. These operations were small, discrete and self-contained once launched.

This changed with Operation TORCH in North Africa when 1 Airborne Brigade was employed as a component of the larger invasion force. While 1st Airborne Division in Britain retained full command of the formation, it was placed under the operational command of the Allied Forces HQ (AFHQ) in Tunisia. This pattern was repeated throughout the rest of the war. Airborne formations were under full command of an airborne HQ for training and administration but were then placed under the operational command of a conventional field HQ for the duration of an operation. In principle the highest HQ in the field exercised operational command of an airborne force but it was often delegated to a tactical HQ to enable closer cooperation with the ground force.

Devolving command of an airborne force to the tactical level required careful consideration. Few if any senior tactical commanders had any depth of experience of handling airborne units: Horrocks, as commander of XXX Corps, had not even seen a parachute drop before Operation MARKET GARDEN.[8] A tactical commander had to be given clear direction in order to ensure that airborne forces under his command were employed within the overall planned scheme of operations. Despite his rank and the large army under his command, General Sir Miles Dempsey, the commander of British Second Army, was still operating at the tactical level with little influence over the air forces required to execute any airborne operation. He had British airborne forces placed under his command during D-Day, Operation MARKET GARDEN and Operation VARSITY in support of the Rhine Crossing in March 1945. He had very clear views on the utility and employment of airborne forces during the invasion of Normandy and ensuing operations. His initial plan for Operation OVERLORD envisioned 6th Airborne Division being placed under command of I Corps for the landings in Normandy. He then foresaw 1st Airborne Division being placed under the command of VIII Corps for the subsequent breakout. Dempsey believed there were two possible methods of deploying 6th Airborne Division east of the River Orne during the first phase of the operation. One option was to land the force directly behind the beaches at Franceville to allow 1 Special Service (SS) Brigade to land unopposed and achieve its immediate objectives inland, thus securing the main landing's left flank. The second possible method was to use the airborne division to capture the bridges over the Caen Canal and River Orne over which 1 SS Brigade could then pass following a landing on the main beaches. Dempsey selected the latter option because he considered it would take the airborne division too long to neutralise the beach defences at Franceville and the SS Brigade would have to advance too far from the beaches to reach its objectives.[9] The success of the airborne landings on D-Day suggests that his decision was correct.

Dempsey's views on the employment of airborne forces under his tactical command during the breakout from the beachhead were equally unequivocal. Recalling that during the Italian campaign outflanking amphibious operations had successfully turned enemy defensive lines, he concluded that airborne forces in Normandy could execute this outflanking role. Certain principles had to be adhered to. The airborne landing had to be made close enough to the main force to ensure every prospect of a link-up being made before the landing became a liability. The commander of the main body had to be prepared to use maximum force in order to assist the landing. Dempsey also correctly understood that the greatest effect could be achieved by employing airborne forces *en masse*, committing a brigade as a minimum, and preferably a full division.[10]

Dempsey obviously had a deep professional interest in this new form of warfare and made copious personal notes on official papers. He transmitted this enthusiasm to his subordinates and encouraged them to consider the intricacies of airborne operations. He set Lieutenant-General Sir Richard O'Connor, the commander of VIII Corps, a problem concerning the employment of an airborne force during a breakout from the Orne bridgehead. The crux of the matter, as Dempsey noted, was whether the airborne force should be committed once success was ensured, to turn an enemy withdrawal into a rout, or in cooperation with the main force in order to gain success.[11] O'Connor did not commit himself to one or the other of the two conditions but in response to his commander's orders he produced detailed plans outlining the possible methods of employing an airborne division in Normandy. These included control, liaison, fire plans and communication diagrams.[12] Dempsey encouraged his subordinates to liaise closely with the airborne establishment. O'Connor and Lieutenant General John Crocker, the commander I Corps, held planning meetings with Browning, Commander I British Airborne Corps, to consider the problems set by their commander.[13]

There were issues with delegating command and control of airborne operations to lower tactical HQs. As the Combined Chiefs of Staff noted, one result was that the vital and necessary coordination between the Army and RAF was also delegated to the tactical commander. This arrangement was likely to be less successful than retaining command at a higher level, since in order to achieve coordination decisions had to be made by a commander who was in direct control of all the forces concerned, both ground and air.[14] Working at the tactical level meant that a commander could only expect to have airborne formations under his control for discrete operations. For Dempsey there was six-month gap between Arnhem and his next chance to deploy airborne troops during Operation PLUNDER, the Rhine Crossing in March 1945. Corps commanders such as Crocker and O'Connor would have even less regular opportunities. It was therefore difficult for the commander to build consistent experience in deploying airborne forces.

At the operational level of command, what contemporarily would have been thought of as the theatre level, the land commander had more opportunity to develop concepts and doctrine for the employment of airborne troops.[15] He would also have a closer and more regular relationship with his air force counterparts. He could have airborne formations under command for protracted periods or even throughout the entirety of a campaign. This might include very different types of operation in various terrain and conditions. Alexander commanded at this level, having taken over as C-in-C in the Middle East in August 1942, but had no direct influence over the British airborne operations during Operation TORCH in November 1942. He took command of Eighteenth Army Group in Tunisia in February 1943, and as General Dwight Eisenhower's deputy in command of all land operations in North Africa he continued to work at the operational level. He was thus in command for the planning and execution of the invasion of Sicily, Operation HUSKY, which included 1st British Airborne Division. During HUSKY three British airborne operations were planned with two being executed. Alexander retained command of the Division during the Italian campaign until October 1943, when it was withdrawn leaving 2 Independent Parachute Brigade in the theatre until this was committed to the invasion of Southern France, Operation DRAGOON, in June 1944. While under Alexander's command the Brigade conducted only one minor airborne operation to harass the Germans in Italy, Operation HASTY from 1 to 7 June 1944. The Brigade returned to Italy in January 1945 and between 6 March and 4 May thirty-two airborne operations were initiated then cancelled by Alexander's HQ. At one point the Brigade HQ was simultaneously planning eleven separate operations, all of which were due to take place within a fortnight, none of which were executed.[16] This demonstrated a lack of foresight and the inability to clearly identify the best opportunity for the employment of the airborne force. Apart from DRAGOON and the brief interlude of HASTY, Alexander's British airborne forces were employed as line infantry in Italy, causing a steady casualty rate, including Major General Hopkinson, the GOC 1st Airborne Division.

The Combined Chiefs of Staff indirectly criticised Alexander for his lack of positive action and misuse of his airborne forces on two separate occasions. In March 1944 they noted, 'since airborne units are designed primarily for use when opportunity arises, they should not be employed in roles for which other units are suitable. In the past some commanders have been guilty of violation of this principle and have failed to appreciate the proper role of airborne units'.[17] Because Alexander employed 2 Independent Parachute Brigade as conventional infantry it made it far more difficult to take advantage of any opportunity more suited to their specialist role. In August 1944 Alexander was more obviously the object of criticism as General George C. Marshall, Chairman of the Combined Chiefs of Staff Committee, stated there had been no airborne operations in Italy since Operation AVALANCHE at Salerno in September 1943. He felt that this demonstrated a lack of understanding or even misuse of airborne forces.[18]

Despite his experience Alexander left little evidence of having conducted any deep analysis into the most effective employment of his allotted airborne forces.[19] Despite Marshall's assessment he did clearly recognise the value of airborne forces. Following Sicily he reported to the COS Committee on Operation HUSKY, praising the achievements of 1st Airborne Division and emphasising the advantage gained by its employment during the operation. He believed that it was time to make the expansion of the airborne corps a high priority and listed what needed to be done in terms of supplying equipment and improving training.[20] Shortly after the invasion of mainland Italy Alexander wrote to CIGS outlining his intention for the forthcoming campaign: to advance north and knock Italy out of the war. His concept involved Fifteenth Army Group moving steadily up the peninsula, bound by bound. He intended to take every opportunity to outflank the enemy defensive lines by amphibious assault and, with the allies' air superiority, by airborne assault. 'It will be possible, and highly desirable, to employ a portion of our two airborne divisions to secure key points in the rear of the enemy by attacking him over the open vertical flank.'[21]

Alexander liked the term 'vertical flank' and used it on several occasions. He described it again in a letter sent to Brooke in autumn 1945 concerning the future value of airborne forces.[22] It therefore appears anomalous that he did not employ his airborne formations to greater effect during the Italian campaign. Admittedly there were limited airfields in Sicily and southern Italy that were sufficiently developed for medium and heavy aircraft, and where they did exist bombers took priority over transport aircraft. He did use 82nd United States Airborne Division in the parachute role during the operations in the Salerno beachhead but only after he was urged to do so by General Walter Bedell Smith, Eisenhower's Chief of Staff.[23] 1st British Airborne Division and later 2 Independent Parachute Brigade remained largely confined to conventional infantry operations.[24] Alexander therefore represents something of a paradox when it comes to his handling of airborne forces. He clearly recognised their value but did not seem able to identify appropriate operations for their employment. Identifying opportunities that had a good probability of success required perception and timely decision-making from a commander. Alexander was not famed for his intellect: one officer commented that he could not imagine the Army Group Commander ever producing a plan, let alone a good plan.[25] The command of airborne forces at the operational level required the ability to recognise the unique advantages that the airborne capability represented and then be able to overlay these onto the overall plan, so as to realise those advantages without prejudicing the main effort. General Sir Bernard Montgomery, Commander Eighth Army, and as such Alexander's subordinate, had no doubt that he did not possess these qualities.

> In my opinion 15 Army Group is a very bad and inefficient H.Q. I think the staff there works under great difficulties since they find it quite impossible to get any decision out of ALEXANDER, or any firm line of country on which to work. There is no proper planning or thinking ahead. ALEXANDER does not really know clearly what he wants; and he has very little idea how to operate the Armies in the field.... He does not understand the offensive and mobile battle; he cannot make up his mind and give quick decisions; he cannot snap out clear and concise orders. He does not think and plan ahead.... The whole truth of the matter is that ALEXANDER has got a definitely limited brain and does not understand the business; the use of air power in the land battle is a closed book to him.[26]

Montgomery, along with Eisenhower had, by the end of the war, built up a greater depth of experience in handling airborne forces at the operational level than any other senior commander. From North Africa, through Sicily, into Normandy and across the Rhine these two officers planned many and executed several medium and large-scale airborne operations involving British parachute and airlanding formations. Montgomery and Eisenhower's approaches to airborne warfare often overlapped if they were not entirely coincident.

Eisenhower was given command of the Allied landings in North Africa on 26 July 1942. His first outline draft plan for TORCH, written at the HQ European Theatre of Operations on 9 August 1942, made no mention of employing airborne troops. Neither did the next iteration written on 21 August 1942 and delivered to the COS Committee the following day.[27] At around the same time the first thoughts on including an airborne element appeared in correspondence between Major General Mark Clark and Eisenhower. Planning began, and at the beginning of September an Airborne Task Force was established and combined training began. On 15 September Browning was brought in to discuss the use of British airborne troops and their joint training with American air transport. The prominence of airborne forces within the plan began to rise rapidly. By 19 September it had been decided that paratroops were to be used 'as extensively as possible'. Four days later it was considered that the 'use of paratroops is extremely vital to the operation'. It is possible that Browning's introduction to the planning process was instrumental in this escalation as he pointed out to Clark that he was eager to have his whole command included in the operation.[28]

Eisenhower's dispatch from the North African campaign described the airborne operations that took place but drew no conclusions concerning their utility and lessons for the future.[29] There appears to be no evidence that he had thought through any personal concepts for the employment of airborne forces prior to TORCH. Montgomery, in contrast, had dedicated a considerable amount of thought to the application of this new style of warfare well before he arrived to take over Eighth Army in Egypt in August 1942. This is perhaps not surprising as his Chief of Staff, Major General Sir Francis de Guingand noted, 'his [Montgomery's] knowledge of the military art, in all its spheres at the outbreak of war, was, I think, quite unique.... He could talk with exact and technical knowledge of either the sappers' or the gunners' work; as easily as he could discuss that of his own arm's rôle in battle – i.e. the infantry.'[30] It is therefore logical that Montgomery applied equal thought to the concept of airborne warfare when it first emerged. In December 1940, as the commander of V Corps, Montgomery conducted two very ambitious exercises across southern England, which included paratroops in his friendly forces order of battle. During the two short exercises there were four drops involving a total of 250 paratroops. The drops were integrated within the entirety of the exercise activity including air attack, the fire-plan, reconnaissance and bridging tasks. The number employed represented almost the whole of Britain's airborne capability at that time. The exercises were considered a success and helped to convince Brooke of the utility of airborne forces.[31] Montgomery continued to include a proportion of airborne troops in many of his large-scale exercises such as MOREBINGE in August 1941 when training XII Corps and BUMPER in September 1941, a Home Forces exercise for which he was made Chief Umpire.

Having experimented and examined the subject, Montgomery felt confident enough to expose his thoughts on the employment of parachute troops publicly during an address to 500 officers of 3rd and 4th Divisions in March 1941. He expounded his views on the basic concept for the conduct of airborne warfare. These included the inviolable principle that once dropped paratroops should have a good and reasonable chance of linking up with ground forces. They should not be inserted too deep, as any failure to adhere to this would obviously have an adverse affect on morale. From this principle he deduced that paratroops would not be suitable as part of a defensive plan. He considered that harassing tasks and seizing key points ahead of an advancing force presented the most appropriate opportunities to employ airborne troops. He also emphasised the importance of good communications and intelligence in airborne operations. Montgomery was already demonstrating his belief that the place for airborne forces was in pre-planned, deliberate, set-piece operations. 'The employment of parachute troops must have a very definite place in any plan of battle. For this reason I wish greater attention paid to the subject.' His thoughts on the subject were far more advanced than most of his peer group, many of who remained unconvinced by the utility of the new capability.[32]

The first time Eisenhower and Montgomery coincided on the subject of airborne operations was during the planning for Operation HUSKY. The decision at the SYMBOL conference to designate Sicily as the next allied target following victory in North Africa was a turning point for British airborne forces, as Eisenhower decided to use 'airborne troops in the operation on a much larger scale than had yet been attempted in warfare'.[33] Although there was some early dispute between Eisenhower and Montgomery as to the initial dispositions and objectives of the invading forces, the focused use of airborne forces was never in doubt. Both the American Seventh Army and Montgomery's Eighth Army had airborne forces placed under command for the initial assault. 1 Airlanding Brigade was used to seize the approaches to Syracuse including the capture of the Ponte Grande on 10 July 1943. Three days later Montgomery, keen to maintain the initial momentum, used 1 Parachute Brigade to seize the Primasole Bridge over the River Simeto south of Catania.

Despite the high casualty rate, due mainly to the inexperience of the American pilots of Troop Carrier Command and a lack of time for collective training immediately before the operation, both operational commanders agreed that the airborne operations on Sicily had presented important tactical lessons. Montgomery believed that the operation mounted by 1 Airlanding Brigade had accelerated his advance by no less than seven days.[34] However he considered that the operation to gain a firm hold of the Primasole Bridge had been a failure.[35] He had identified an important lesson that confirmed his previous assumptions and his own general approach to operations. Where airborne forces were employed in a pre-planned, deliberate, set-piece operation they could be valuable and stood a good chance of success. When they were used in a hasty operation to exploit success or overcome a tactical impasse the probability of failure was increased. He also developed firm views on the importance of collective training and the importance of dedicated aircrews for airborne operations. He felt it was unfair to commit airborne troops if the air force could not be relied upon to deliver them accurately.[36] The formation of 38 and 46 Groups RAF as dedicated airborne support formations followed the experiences of HUSKY. His observations in Sicily may also explain the fervour applied to airborne training prior to D-Day and the resources made available to it.

Eisenhower believed that 'the outstanding tactical lesson of the [Sicily] campaign was, for me, at least, the potentialities of airborne operations'.[37] He had also drafted a number of lessons to be carried forward to future operations. He shared Montgomery's view that the principal utility of airborne forces was to be found in deliberate operations with a long lead-time to allow the maximum period for coordination between all services and arms. He believed that airborne forces should only be committed to missions suited to their role.

> The force commander's decision to use them must be made only after he is positive that the mission cannot be accomplished by other means more economical or equally suited to the mission. In weighing the decision it must be recognized that airborne operations are both hazardous and difficult of coordination, and can be justified only by a situation which clearly shows the use of such troops to be imperative.[38]

Eisenhower came under pressure to abandon these principles when his chief, Marshall, tried to influence his operational plans from Washington. In February 1944 Marshall was imploring Eisenhower to include an airborne operation deep in the enemy's rear during Operation OVERLORD. He identified the area around Évreux, forty-five miles from Paris, as a possible objective as he believed it would have presented a strategic concern to German forces by threatening the crossings over the Seine. Montgomery believed the plan was mad and criminal in its indifference to the lives of Allied airborne soldiers. Eisenhower was persuaded by Montgomery amongst others to reject the proposal.[39] Eisenhower reported to Marshall that to employ airborne forces so deep at such an early stage of the operation was impossible, as he could not guarantee an expeditious link-up. Marshall felt that without the deep employment of concentrated airborne forces the existing plan for OVERLORD lacked the element of surprise. Eisenhower disagreed but conceded that the use of a large airborne force at some later stage of the operation might be desirable.[40]

After the invasion of Normandy had been consolidated and the breakout had begun the Combined Chiefs of Staff criticised Eisenhower's handling of his airborne forces once again. The Joint Staff Mission in Washington reported to the COS Committee:

> He [Marshall] felt that neither Eisenhower nor [General Sir Henry Maitland] Wilson [C-in-C Mediterranean] had used their airborne potentialities to the full and were inclined to "save them up" and even

misuse them as ordinary infantry.... He felt strongly that the introduction of these large forces in France at the right time and place had very great potentialities and he had no indication of General Eisenhower's plans for their use.[41]

Marshall made suggestions for the future employment of the available airborne formations. He believed they could be dropped on Rouen if German resistance between Caen and Vire weakened or south of Dunkirk as a potential method of eliminating the flying bomb threat. He admitted that these suggestions were the result of only a relatively superficial appreciation and that no serious study of the forces required had been made. Both schemes violated the principles identified by Eisenhower and Montgomery in the past. The Rouen option relied on reacting quickly to an enemy reverse and so it could not have been pre-planned and deliberate to any great degree. The Dunkirk plan stood very little chance of any link-up being made by a land advance before the airborne force could be destroyed in detail.

By this time Montgomery had already been forced to cancel one airborne operation. He had planned to drop 1st Airborne Division around Evrecy on 13 June in order to form the linchpin for an encirclement of Caen. Air Marshal Sir Trafford Leigh-Mallory, C-in-C Allied Expeditionary Air Force had objected to the plan as he thought the scheme would have been expensive in terms of aircraft and casualties and thus prejudice future airborne operations. Montgomery thought Leigh-Mallory 'a gutless bugger' but allowed the concerns of his air commander to influence his operational plan, and on the morning of 13 June he cancelled Operation WILD OATS.[42] Following this incident Montgomery's plans for airborne operations were continually frustrated. Operation BENEFICIARY, an airborne assault to capture the defences around St. Malo; SWORDHILT, an effort to destroy the Morlaix viaduct; and HANDS-UP, a plan to capture the port of Quiberon to assist the American breakout were all cancelled as the ground advance reached the objectives before an airborne assault could be launched. TRANSFIGURE, BOXER, AXEHEAD, LINNET, INFATUATE and COMET suffered the same fate.[43]

It was against this background that Montgomery was given priority in September 1944 to try breaking the static situation on the Belgian/Dutch border and attempt a narrow thrust into Germany to encircle the Ruhr. At the same time Eisenhower was under pressure from Marshall to use his airborne forces. Eisenhower responded by looking at all potential employment such that his staff appeared to have 'decided to buy an airborne product and were shopping around'.[44] So both Eisenhower and Montgomery were induced to abandon the principles for the employment of airborne forces that they had previously expressed. Operation MARKET planned to seize key crossings over the Maas, Waal and Rhine in order to allow an advance by Second British Army to penetrate Germany. Only seven days were available to plan the largest airborne operation of the war thus far, hence the schemes for two previous operations, LINNET and COMET, were stitched together to speed up the process. The depth of the insertion of 1st British Airborne Division behind the German front line reduced the chance of a link-up with Dempsey's army.

Communications at the operational level were poor, intelligence was ignored and the Air Force appeared to dictate crucial parts of the plan. Operation MARKET was thus an aberration in both Eisenhower's and Montgomery's approach to airborne warfare. Although deviating from their principles did not automatically lead to failure it did reduce the chances of success and meant that relatively minor tactical errors were more likely to jeopardise the entire operation.

Comprehensive lessons were drawn from Operation MARKET at the tactical level and it would appear that they were learned at operational level also. The next and final time a major airborne operation was launched in Europe, Eisenhower and Montgomery's principles were fully applied once again. Operation VARSITY, as part of the wider Operation PLUNDER, was a deliberate set-piece battle to cross the Rhine. Montgomery returned to his *forté*, and to ensure that a link up with the airborne force was achieved it was not launched until the ground forces were already across the river. The operation was highly successful despite the casualty rate and all the airborne objectives were achieved within hours.[45] Montgomery acknowledged his experiences when he wrote to Brooke during the autumn of 1945 to give his views on the value of airborne forces. He emphasised that the vital role of airborne forces was most particularly suited to deliberate operations. He could relate from experience that during highly mobile operations the value of airborne forces was limited owing to the considerable amount of time required to plan for their deployment and use.[46]

It has been argued that the long period of readiness between Dunkirk and D-Day provided the opportunity for training and large-scale exercises and thus officers selected for high command had experience leading large forces in mobile exercises such as they had never had before.[47] Numerous airborne exercises were conducted during this period but it would appear that few senior commanders took the chances proffered to experiment with the benefits and limitations of this new opportunity. Those who did, like Montgomery, quickly built up firm ideas concerning their effective employment. The thoughts and enthusiasm of a single commander such as Montgomery could then be effectively communicated to subordinate commanders through the chain of command: thus Dempsey passed ideas down to Crocker and O'Connor.

It had been recognised before an airborne formation had ever been committed to operations that command and control would be a complicated matter.[48] Command of airborne forces could only be delegated to the tactical level for defined periods and limited operations, as the advantages of close cooperation with ground forces could be negated by the inability to coordinate effectively with the air component. The successful handling of airborne forces at the operational level required foresight and understanding to identify opportunities and timely and effective decision making to ensure that there was time to properly plan and coordinate their deployment. Where this was not apparent, as with Alexander, they were misemployed and hence opportunities were lost. Those who did have the opportunity to use airborne forces on multiple occasions over a protracted period of time drew firm conclusions as to their effective employment. These essentially centred on set piece, deliberate operations where the chance of linking up with ground forces was practically guaranteed. Where these principles were forgotten or ignored by operational commanders it made success at the tactical level very difficult.

The Selection and Competence of Airborne Commanders
Operational and higher tactical command of airborne forces required an understanding of concepts, the ability to adapt and apply doctrine and the coordination of ground and air components in order to fit a desired plan. The tactical command of airborne forces at the divisional level and below required different qualities. The unique capabilities, limitations and the technical intricacies of mounting airborne operations needed officers in command who were intimate with the tactics, techniques and procedures associated with airborne warfare. Tactical airborne commanders held the potential to influence the formations and units under their command to a greater extent than the commanders of conventional formations and units. Even the most informed and adept superior commanders were by no means experts in this field. They had not served in airborne units themselves and were therefore less inclined to meddle at the tactical level. They were not subject to the same level of tactical suggestion and prescription as infantry and armoured commanders and their staff, who were continually confronted with Army Training Memoranda and Military Training Pamphlets. During operations, airborne formations were often isolated, deploying in advance of ground formations. They often fought alone for a period of time, without the requirement to conform to flanking formations, and therefore tactics could be dictated by necessity rather than by template. Finally, as will be discussed, airborne doctrine for both training and operations was thin during the war. All these factors left an airborne commander with wider latitude to dictate the manner in which his unit or formation trained and operated.

The initial intakes of airborne commanders had no more experience of airborne warfare than the men they were expected to lead. However, many of these young officers brought with them a wealth of general military experience from various theatres across the world. Frost had served with Assyrian Levies in Iraq before volunteering for parachute duties, Crookenden in Egypt, Palestine and Ulster. Some of them, such as Pearson, Deane-Drummond and Crookenden brought valuable lessons with them from service with the BEF prior to and during Dunkirk.[49] These men were pre-war professional soldiers or had at least served with the Territorial Army. They were young, capable and adventurous and many demonstrated a streak of unconventionality in their early careers. They were the ideal officers to take forward the initial establishment and expansion of Britain's airborne forces. Their limited time in the Army and service in the colonies and on the continent meant they were unlikely to have been irreversibly imbued with conventional military thought. This was important, as airborne warfare required a different, if not unconventional, approach to tactical command.

A commander would have to lead airborne soldiers who had been trained to think and act on their own initiative. On landing he might have to rapidly reconfigure his troops to achieve his objectives. His unit would have to generate speed during the assault, which was essential to compensate for the lightweight nature of their arms, in order to produce sufficient momentum of force to carry the objective. He might have to fight with open flanks and an unguarded rear. These factors required a style of command that was not overly prescriptive, relied on decentralisation and allowed men and officers to achieve their objectives through their own means rather than by dogmatically following detailed orders. This approach required a high degree of trust, mutual understanding and good communication. This was more akin to the German style of command, which they named *Auftragstaktik* and was latterly known as 'mission command' by the British. A commander set out his intent for a battle and subordinates were given wide latitude to act as they saw fit as long as that intent was achieved.[50]

This approach to command was certainly recognised within the airborne community and the chain of command and training were adapted to bring forward its advantages. Lieutenant Colonel James Hill recalled the effects of an acceptance of 'mission command' following Operation TORCH where he commanded 1 Parachute Battalion.

> Fortunately throughout our training we had always appreciated that it might be necessary for parachute troops to carry out an operation quickly under great difficulties at very short notice, and a system of decentralization had been very carefully worked out.... If we had not had this [approach] our new task would have been totally impossible.[51]

Certainly in many cases subordinates recognised and appreciated this style of command in their superiors. Crookenden considered that his brigade commander, Brigadier The Honourable Hugh Kindersley, had these qualities. 'I think he will be v[ery] good as a Com[man]d[er], since he has the gift of leaving people to get on with it.'[52]

Another symptom of the traditional British bureaucratic attitude to command could be seen in an over-reliance on copious written orders. Detailed and cumbersome, over-prescriptive orders reflected the Army's hierarchical ethos and structure but were not responsive enough to be of use during a highly fluid battle.[53] Once again Britain's airborne forces appear to have recognised this weakness and modified procedures, attempting to keep written orders brief and leaving subordinates scope for autonomy. The entire planning instructions issued by HQ 6th Airborne Division prior to the invasion of Normandy ran to just approximately 2,500 words. The instructions gave each subordinate commander clear objectives, task and constraints but left him with the freedom to configure and deploy his formation as necessary in order to be most effective. Only where objectives were critical to the overall D-Day landings, such as the battery at Merville and the bridges over the Orne and Caen Canal, did the orders become more prescriptive.[54] This style of instruction gave a subordinate commander the chance to think through the tasks he had been given and conduct his own appreciation of the situation and the approach required rather than simply following superior orders. This can be seen in the appreciation written by Hill, as commander of 3 Parachute Brigade to interpret his D-Day instructions disseminated by HQ 6th Airborne Division.[55] This method allowed a commander to build in flexibility to cope with the inevitable last minute changes required in most airborne operations once troops had landed and the force available could be assessed. It allowed subordinates to use their initiative if the original plan became invalid. According to Gale, initiative was the primary quality of an airborne leader.[56] As long as the commander's intent was understood and objectives were accomplished the method and means were less important. Where a commander was over-prescriptive and exercised too direct control over his subordinates an airborne operation could quickly go awry, as will be seen.

The promotion and re-assignment of qualified airborne officers was not always sufficient to fill the new appointments created by the surges in expansion of the establishment, particularly between late 1941 and late 1943. The formation of 1 Parachute Brigade during the second half of 1941 necessitated the simultaneous appointment of three battalion commanders and a brigade commander. These selections required careful consideration as the commanders in such a small and still relatively experimental establishment could have an effect out of all proportion to their position. The command of a parachute battalion was considered unique and required first-class, independently minded officers to fulfil it.[57] During this period a brigade commander was given the opportunity to select his own battalion commanders and their company commanders.⁵⁸ Gale was selected to command the new 1 Parachute Brigade. There appears to be no evidence as to why Gale was singled out for this appointment. To that date he had had a relatively conventional career: a machine gun officer during the First World War where he won the MC, he attended Staff College at Quetta, India between the wars and was working as a staff officer in the War Office for the first two years of the Second World War. During this period it is possible that someone in authority identified him as having the necessary qualities to command a parachute brigade.⁵⁹ He then selected his battalion commanders as Lieutenant Colonel Ernest Down, who already commanded 11 SAS Battalion, Lieutenant Colonel Edward Flavell who had been Gale's company commander during the previous war and Lieutenant Colonels Lathbury and Hope-Thompson.

As the establishment expanded to a division during 1942 and then to two divisions during 1943 some of these officers and their peers made steady progress through the ranks. Down was promoted twice and commanded 2 Parachute Brigade and then 1st Airborne Division. Lathbury, after a spell working for the Director of Air in the War Office, was promoted and commanded 1 Parachute Brigade as Hill became the commander of 3 Parachute Brigade. Gale moved from brigade command to become the newly designated Director of Air and was then promoted to command 6th Airborne Division, subsequently being appointed commander of I British Airborne Corps. Even so, it was still necessary to bring officers in from outside the airborne fraternity in order to fill all the new appointments. These included officers such as Brigadier Nigel Poett who commanded 5 Parachute Brigade in Normandy, and on the Rhine, Brigadier 'Shan' Hackett, commander of 4 Parachute Brigade at Arnhem, and Major General Eric Bols, GOC of 6th Airborne Division during the Rhine crossing.⁶⁰ These officers brought with them a wide variety of experience and hence each had a different approach to their new commands. For most there is no evidence whatsoever to suggest why they were selected to command what was considered a *corps d'elite* beyond the fact that they had obviously demonstrated a certain level of competence in their careers thus far.⁶¹ Hackett was not even an infantryman. One of the factors that connects these commanders' airborne careers is the mystery of their selection for the appointment. For some there is evidence of patronage. Certainly Bols was personally selected to command 6th Airborne Division by Montgomery and was informed of the fact as the Field Marshal pinned his DSO on his chest.⁶² Although the methods by which these officers were selected for airborne command are unclear, on the whole they appear to have been successful and performed equally well as their indigenous peers.

The officers selected as the original commanders of the airlanding brigades had both been amateur civil pilots before the war. Colonel

Chatterton, the Commander Glider Pilots, considered this was an unhealthy precedent on the principle that a little knowledge is a dangerous thing.[63] Kindersley assumed command of 6 Airlanding Brigade in May 1943 having commanded a battalion of the Scots Guards. His subordinate officers considered him pleasant, efficient and invaluable in battle.[64] Brigadier Frederick Hopkinson preceded Kindersley, having taken command of 1 Airlanding Brigade in October 1941. The Brigade had just been created from the conversion of 31 Independent Brigade Group and Hopkinson took over from that formation's commander. Hopkinson had served in the First World War and had made a name for himself in 1940 with the BEF as the commander of the GHQ Reconnaissance Unit, later to develop into the Phantom Group.[65] Deane-Drummond found Hopkinson dogmatic and inflexible while Chatterton considered him, 'an unusual little man with a little knowledge of the air… he was very ambitious'.[66] It was possibly this latter trait that motivated his actions after he was promoted to commander of 1st Airborne Division in April 1943. During the planning for Operation HUSKY Hopkinson bypassed the chain of command and approached Montgomery personally in order to try to ensure the participation of his Division during the invasion of Sicily.

Despite having commanded an airlanding brigade for eighteen months Hopkinson's appreciation of the requirements of a mass airborne landing was low. The airborne collective training establishment in Britain had yet to reach the stage where brigade level exercises could be practised. The training available to the glider pilots, in particular in North Africa, prior to HUSKY was insufficient to prepare them adequately for the operation. Hopkinson was aware of this as Chatterton had advised him as such and recommended cancelling or at least reducing the airborne participation in the operation. Hopkinson threatened Chatterton with the sack if he did not comply.[67] Browning, Hopkinson's superior within the airborne establishment, was also concerned and wanted to discuss the operation with him. Hopkinson deliberately evaded Browning by missing a pre-arranged meeting in Algiers in order to avoid reconsidering his plan.[68] The result of Hopkinson's over-enthusiasm in terms of airborne casualties has already been discussed. After HUSKY Hopkinson landed with his division by sea at Taranto and accepted the Italian surrender of that city. Shortly after he was killed in action while observing one of his forward units.[69]

Throughout the war the airborne establishment was commanded at all levels by a mixture of officers who had worked their way through the airborne ranks and those with external, conventional experience brought in at various levels to take command when required. At first glance it may appear that bringing officers in from other parts of the Army could have stifled the careers of those looking for promotion within. Some very capable airborne officers appear to have had their careers truncated. However, in most cases there were very good reasons for this. Lathbury, having commanded a brigade in Sicily and at Arnhem, would have been a candidate to take over 6th Airborne Division from Gale in December 1944 had he not been wounded and captured during Operation MARKET GARDEN.[70] Frost, one of the most experienced British airborne officers having commanded at Bruneval, in North Africa, Sicily and at Arnhem did not make it beyond Lieutenant Colonel during the War. He might well have been selected to succeed Lathbury as brigade commander had he not also been wounded and captured at Arnhem. In other cases the reasons for a career decelerating were less obvious.

Having been appointed commander of 1st Airborne Division, Down was moved to a much lower profile theatre to take command of the newly formed 44th Indian Airborne Division. It is probable that this was a political appointment designed to placate the Indian General Staff and convince them that the War Office was taking its airborne aspirations seriously. Down was a victim of his own success. He was considered by one of his biographers to be the toughest of all the Second World War airborne generals, hard but held in high regard by those he commanded.[71] At least one of his subordinates held him in very high opinion. Deane-Drummond found that he set high standards and was ruthless in applying them. He considered Down was the best airborne theorist of the war, who had taken time to study the problems associated with landing deep into enemy territory.[72] Down was known for being taciturn, a quality displayed by his assessment of 154 Parachute Battalion. Having observed its training during an exercise in India in October 1944 he commented, 'The jumping with rifles and LMG [light machine gun] carried down on the man was good, the accuracy of dropping bad. The work of the Co[mpan]y on the ground average.'[73]

From the end of 1941 and throughout 1942 the Indian Viceroy, Linlithgow, the Secretary of State for India, Amery and C-in-C India, Wavell had all lobbied the War Office to little avail to increase the rate of development of their own airborne force. In late 1943 Browning visited India and recommended that more effort be put into the Indian airborne establishment. By dispatching the European theatre's most competent and widely respected airborne soldier Whitehall could convince India that it was taking its aspirations seriously without having to commit significant resources. Down left for India in January 1944 and was clearly not pleased at having to give up command of 1st Airborne Division.[74] 44th Indian Airborne Division was never committed to battle as an entire formation and so Down's qualities were never tested at that level. He handed over 1st Airborne Division to Major General Robert 'Roy' Urquhart who subsequently commanded at Arnhem.

It has been generally accepted that Urquhart, like Bols, was personally selected for this command by Montgomery.[75] Montgomery certainly knew Urquhart before the appointment. He was commissioned into the Highland

Light Infantry in 1920 and served in Malta and India before the war. In late 1940 Urquhart was posted onto the staff of 3rd Division and was subsequently made the chief administrative officer. It is possible that he was first noticed by Montgomery, under whose command he was serving during this period. Between 1941 and 1944 he commanded a battalion of the Duke of Cornwall's Light Infantry, served as a staff officer with 51st Highland Division and XII Corps and he commanded 231 (Malta) Infantry Brigade Group. During the majority of this period Urquhart was subordinate to Montgomery at some level. Montgomery put great store in personally selecting many of his junior commanders and estimated that he spent a third of his working hours considering new appointments and officers' performance. He deemed the most important qualities of a leader to be initiative, drive, determination, moral courage, resolution, and the ability to radiate and inspire confidence.[76]

Montgomery favoured commanders with a proven combat record. However, between the outbreak of war and January 1944 Urquhart had only commanded 231 Brigade in action and for just approximately twenty-four days on Sicily and for a further three days in Italy.[77] The Brigade had an independent role during the invasion of Sicily under Montgomery's Eighth Army. One of the most difficult actions carried out by the Brigade was the capture of a crossing over the River Dittaino conducted during the night of 18 July 1943. The battle was recorded by the brigade history as 'a perfect little set-piece of military manoeuvre'.[78]

Urquhart the divisional commander was a product of his experience, however, and the understanding accrued from conventional operations was not fully transferable to the airborne method of operating. Commanders arriving as newcomers to the airborne establishment might be inclined to think that it represented just another means of arriving in battle, ignoring the unique physical, mental and tactical challenges.[79] Urquhart had ample time prior to taking his division into battle to acquaint himself with the distinctions. When he assumed command of the division in Lincolnshire it had only recently returned from the Mediterranean where it had been committed to operations for an average of seven months, twelve months in the case of 1 Parachute Brigade.

Formations and units that have spent protracted periods on operations more often than not require intense confirmatory training in order to bring everyone back to a common base-line, remove bad tactical habits acquired while deployed and prepare for subsequent operations. Urquhart was well aware of this and knew that his main task prior to the projected date for Operation OVERLORD was to concentrate on brigade and divisional level training.[80] It has to be assumed that the divisional commander also benefited from this extended period of training. There is little evidence available as to Urquhart's thoughts and methods during this period but by June 1944 he might reasonably have acquainted himself with the idiosyncrasies and particular requirements of airborne warfare and any changes to his own command style that he might have needed to affect in order to accommodate them.

The course and results of 1st British Airborne Division's battle at Arnhem have been extensively documented but Urquhart's role as commander has seldom been subject to any depth of critical examination. His contribution to the planning process for Operation MARKET produces mixed opinions, particularly the selection of the drop and landing zones eight miles west of the critical road bridge across the Rhine. Urquhart believed that the air force had to remain responsible for all planning up to the point at which paratroops were dropped or gliders were released.[81] Despite this approach Urquhart's biographer maintains that he did personally select the drop zones.[82] One divisional staff officer is certain that the commander took no part in the process and that the selection was left to the RAF and Troop Carrier Command.[83] Certainly Urquhart was offered alternatives, including the possibility of a *coup de main* glider landing against the main road bridge as had been achieved, albeit on a smaller scale, during the early hours of D-Day.[84] Whether Urquhart was part of the decision-making process or not, the Division landed on DZs and LZs some eight miles from its main objective. However this in itself was not an entirely calamitous situation. There may even have been advantages to this position as it allowed the units to form up almost entirely unmolested by the enemy. The distance did not make seizing the main bridge impossible, as Frost's 2 Parachute Battalion was to prove. It was the plan and conduct of the approach and assault from the DZs and LZs into Arnhem that betray Urquhart's unfamiliarity with airborne warfare and the approach to command it required.

The plan for the Division post the initial landing involved the three parachute battalions of 1 Parachute Brigade advancing along three separate routes into Arnhem to seize the crossings over the Rhine, while the three airlanding battalions of 1 Airlanding Brigade remained on the DZs and LZs to defend them for subsequent drops and landings.[85] However the plan for the advance began to unravel early in the operation. The divisional reconnaissance squadron had been tasked to race ahead of 1 Parachute Brigade to seize the main road bridge but its jeeps were quickly halted by fierce resistance. The surviving vehicles returned to their start point and became inactive for a prolonged period, apparently impotent without new orders. Without reconnaissance and lightly armed, the lead parachute battalions were repeatedly halted by relatively minor action which required counter-attack before they could be bypassed. This intermittent progress made it difficult for the battalions to generate the speed of advance they required to overcome the disadvantages of their light armament.

With the initial advance not moving as quickly as planned Urquhart felt he had to intervene personally. He left his divisional HQ and moved forward to try to expedite 3 Parachute Battalion's progress along the centre route through Oosterbeek. When Urquhart arrived at Lieutenant Colonel Fitch's Battalion HQ he found the brigade commander, Lathbury, already in attendance. The Battalion had been held up by relatively light but determined opposition from 16 SS Panzer Grenadier Depot and Reserve Battalion, which had been exercising in the area when the drop took place. Rather than bypassing the enemy and continuing to the bridge Fitch spent time trying to eliminate the enemy unit. By dusk 3 Parachute Battalion was still engaged on the centre route, several miles short of the bridge. Urquhart, with Lathbury and Fitch, took the decision to take up all round defence for the night, disengage from the enemy before first light and move on to the southern route to continue towards the bridge the next day. It seems remarkable that, having come forward to try to speed things up, Urquhart was then content to allow this overnight delay and inevitable loss of momentum.[86] The ineffectiveness of the reconnaissance squadron after an initial setback and the inability of parts of 1 Parachute Brigade to bypass the enemy and make best speed to the bridge are not indications of a division that fully embraced a 'mission command' ethos. Urquhart's decision to leave his HQ and then halt 3 Parachute Battalion for the night betray a lack of understanding of the situation and, more fundamentally, the essence of airborne warfare and the absolute requirement for speed in the attack.

By the next morning the Germans had moved reinforcements from 9th SS Panzer Division to form an impenetrable blocking line to the west of Arnhem. Throughout the remainder of the battle no further significant advance could be made towards the bridge and those already there were steadily destroyed in detail. Certainly the Germans pointed to the decision to halt on the first night as the critical mistake of the battle, as noted in an official report. 'The adversary only made one big mistake, and that not only thwarted his own intentions but exposed him to destruction.... If the enemy had pushed straight on to Arnhem after having surrounded the [16 SS Panzer Grenadier Depot and Reserve] B[attalio]n instead of trying to wipe it out, he would have succeeded in capturing the town.' The report concluded that British airborne forces lacked the capacity for 'courageous and resolute advance'.[87] Urquhart's decision to move forward to personally encourage Fitch is understandable given the breakdown of radio communications shortly after landing. Military doctrine also encouraged this practice, stating that the coordination of the infantry battle relied upon the timely arrival of the divisional commander at a position from which he could make decisions that could directly affect the course of operations.[88] However, this had been written with a conventional infantry formation on a linear battlefield in mind. Moving to forward units during an airborne battle, where they might be isolated or have unprotected flanks was more perilous. By nightfall on the first day of the battle, cut off from his HQ with 3 Parachute Battalion, Urquhart realised that he was losing control of his division.[89] He would not return for forty-eight hours. The Divisional Commander's actions during the early part of the battle have been summarised by one member of his staff as being like 'a wet hen'.[90]

This criticism of Urquhart is specific to that situation and cannot be applied across his career, as he was obviously highly successful as brigade commander in Sicily and Italy. It is not even applicable to the whole of the battle at Arnhem. It was certainly not 'a perfect little set-piece of military manoeuvre' but once the battle became static Urquhart's conduct of the defence was admirable and his planning and execution of the withdrawal across the Rhine was almost impeccable. However, during the planning and critical early phase of the battle Urquhart appears to have failed to recognise and apply some of the basic principles of airborne warfare, primarily the necessity for initiative and independent action through the application of 'mission command', and the critical requirement to maintain speed during the assault. His decisions and actions during these hours had ramifications beyond Operation MARKET GARDEN and had a direct influence on the direction of future British airborne tactics. The shadow that Arnhem cast over Operation VARSITY will be examined in the following chapter.

One officer dominates the subject of British airborne forces during the Second World War. Sometimes described as the 'father' of airborne forces, Lieutenant General Frederick 'Boy' Browning was commissioned into the Grenadier Guards in 1915, won the DSO during the First World War for his actions during the capture of Gauche Wood and was awarded the Croix de Guerre. Between the wars he was stationed in Egypt, at the Guards Depot and as adjutant of the Royal Military College, Sandhurst, where he established the tradition of riding a horse up the steps of Old College during passing out parades. An accomplished sportsman he represented England at hurdling in the Olympic Games and was a member of the British Olympic bobsleigh team. Browning commanded 2 Battalion Grenadier Guards at the outbreak of the Second World War and during the retreat to Dunkirk.[91] He then commanded 128 Infantry Brigade and 24 Guards Brigade in England before being appointed Commander Paratroops and Airborne Troops on 29 October 1941. He took over the new command with the acting rank of Major General the following week.[92]

Many of Browning's character traits are well documented. Immaculately turned out, he was suave and affable and charmed many of those he met within a very short space of time. Chatterton recalls how he 'met 'Boy' Browning for the first time, and fell under his spell as everyone did'.[93] Gale attributed him with a colourful personality, a ready smile, great resolve and determination but with a quick temper.[94] One of Browning's staff officers considered that the enthusiasm with which the early airborne division strove forward and overcame obstacles radiated directly from the commander himself.[95] However not everyone was as complimentary. Pownall, whom Browning succeeded as COS to Mountbatten in Southeast Asia, believed he was rather nervy and highly strung.[96] Browning's supreme self-confidence and fastidious demeanour were interpreted by some as signs of arrogance and Horrocks believed that his immaculate dress led others to underestimate him.[97] This was particularly true of foreign allies and members of the other services who did not necessarily understand that this was the natural manner of a Guards officer. General James Gavin, Commander of the American 82 Airborne Division, found Browning dapper and handsome but believed that his ambition was misdirected and that he lacked experience and understanding of airborne warfare. Their meetings frequently appear to have resulted in antagonism.[98]

Another characteristic that Browning inherited from his Guards background was his obsession with discipline and bearing. In 1942 Browning produced an eight-page pamphlet to be distributed across the airborne establishment entitled 'Discipline – The Only Road to Victory.' In it he explained that the ultimate objective of discipline was to ensure that when a battle was not going to plan, when immediate commanders had been killed and strain and exhaustion were setting in, the soldiers' natural reaction would be to hold on and keep fighting.[99] In October of the same year Browning addressed the officers and men of 1 Parachute Brigade prior to their departure for North Africa on Operation TORCH.

> This war cannot be finally won nor can the intervening battles be fought, without the aid of airborne troops. I have no doubt in my mind that you will furnish the proofs of this, by your deeds, in the months to come. More than this, I have the utmost confidence that by your fighting efficiency, your speed of thought and action, and above all your bearing and discipline, you will most worthily uphold the good name of the airborne division.[100]

Obviously this speech is an attempt to motivate a body of men prior to battle but in it Browning implied that he considered bearing and discipline to be more important individual factors in airborne warfare than speed and initiative.[101]

Browning's approach and attitude and his ability to create friction through his apparent arrogance caused Frost to question his suitability for command of airborne forces. 'Was this the right background for someone who ought to be able to think quite new thoughts and realize that airborne forces would have to rely entirely on air forces?'[102] Certainly there is not much evidence that Browning had many significant 'new thoughts' or if he did they were not committed to paper. As the ultimate commander of such a new and unique force it might be expected that Browning would have had serious ideas concerning the concepts and doctrine for the employment of his airborne troops and methods for training. However, practically no papers of this type appear to have emanated from his HQ when he was divisional and subsequently corps commander. When he was given an opportunity to expound his views on the nature of airborne warfare the results were surreal. At one point during 1942 Browning gave an address to all the officers of the Airborne Division. His speech, lasting approximately twenty minutes, was intended to be on the strategy and tactics of airborne forces. The lecture that he gave, entitled 'A Dream' was quite bizarre. Set at some distant, undefined point in the future he described a huge fleet of vertical take-off, jet aircraft running on a novel fuel source. The aircraft had fully autonomous weapons that could distinguish soldiers from civilians and casualties were 'beamed up' into the aircraft to be treated. The address had absolutely no relevance to the reality of the moment. Quite what impression the officers of the Division, eager to hear their leader's ideas on how the inevitable forthcoming battles might be fought, left with can only be guessed at.[103]

If he was not a doctrinal visionary then Browning did have compensating qualities. He had the ability to move forward projects that, in the hands of others, had apparently stalled. He appears to have been able to produce solutions to previously intractable problems. He was what today might be termed a 'mover and shaker'. Browning was well connected and could present ideas and requirements at the highest level, bypassing bureaucratic obstacles. In May 1942 he presented the Air Ministry's paper on the concept of a 'Nucleus Force' of transport aircraft directly to the Prime Minister, CIGS and CAS. Browning did this because he felt that no one else had placed any constructive ideas before Churchill and that it was an opportunity to offer some insight into the problems facing airborne forces.[104] Two months later the establishment of 38 Wing RAF was doubled.

Browning also provided impetus to events in India. 50 Indian Parachute Brigade had been formed at Delhi in October 1941. However, by mid-1943 not much progress had been made and despite the requests and protests of the Viceroy, Secretary of State and C-in-C India, expansion was painfully slow and the brigade was 1,200 men under strength. Browning was sent to India to report on the situation on the orders of VCIGS, Lieutenant General Sir Archibald Nye in September 1943. He was given a comprehensive list of objectives by the War Office, and following a five-week tour of the country, he returned to London with eight recommendations. These included the formation of an airborne forces depot at Rawalpindi and the raising of an Indian airborne division. The results were immediate. The C-in-C, General Sir Claude Auchinleck, ordered thirty British officers to be posted to 50 Indian Parachute Brigade before Browning had even finished his tour. By December 1943 several units had been expanded and reorganised. In January 1944 Down arrived in India to raise and train 44th Indian Airborne Division.[105]

Browning's penchant for detail on the parade square extended to his staff work. He was willing to immerse himself in the issues which, while they could not have been considered minutiae, were the less glamorous aspects of airborne expansion but important to success nonetheless. He detailed the organisation and duties of the airborne staff; he argued that parachute battalions should retain some of their pre-conversion identities; he requested the provision of smoke bombs that could be dropped alongside paratroops to conceal them on the DZ.[106] Browning, particularly up until the invasion of Sicily, frequently had to battle against War Office indifference and the resistance of the Air Ministry, what one of his staff officers termed the 'many doubters in the higher realms of authority'.[107] But his determination and connections meant that he was eventually successful more often than not.

Gavin had commented on Browning's ambition as being a negative aspect of his character. However the majority of his ambition was genuinely on behalf of the airborne force that he earnestly wished to develop and progress as far as possible. As well as working on those mundane but essential subjects required to ensure success he also took every opportunity to ensure that British airborne forces were considered for deployment on active operations. His inclusion in the planning process for Operation TORCH in September 1942 resulted in a far higher prominence being given to the role to be played by 1 Parachute Brigade.[108] During the planning for Operation RUTTER (later to become Operation JUBILEE), the raid on Dieppe, Browning volunteered the use of parachute troops and pushed hard to have a glider-borne force included.[109] Following the cancellation of the airborne element of that ill-fated operation he lobbied GHQ Home Forces and expounded his view that his airborne forces should be taking part in small raiding operations at the rate of one per fortnight.[110] Despite this obviously over-ambitious assertion he would not, unlike Hopkinson, try to commit his force whatever the consequences. He withdrew the airborne element from a projected raid on Alderney in May 1942, although his reasons were unclear.[111] In May 1943 he advised against the use of paratroops on Operation CORKSCREW, the capture of the island of Pantellaria in the Mediterranean, due to the projected casualty rate.[112] Following an operation, whether it had been fully successful, such as the raid on Bruneval, or if there had been serious set-backs, as with the invasion of Sicily, Browning published a detailed report listing the faults and achievements and actions required to take development forward.[113]

However, there was a streak of personal ambition in Browning, fed in part by vanity, which showed itself during Operation MARKET GARDEN. In mid-1944 Browning was the commander of I British Airborne Corps as a Lieutenant General. There had been considerable inter-allied manoeuvring for control of First Allied Airborne Army, a position that he had coveted but which had been given to an American officer, Lieutenant General Lewis Brereton. Browning had to be content as his deputy. Following D-Day several airborne operations were planned and cancelled and once again inter-allied rivalry had been apparent during this process. Once Operation MARKET GARDEN was confirmed Browning pushed for a corps HQ to be deployed into Holland to control the two American and one British airborne divisions that participated. Acutely aware that he had not been in action as a senior commander Browning volunteered his own corps HQ.[114] The HQ of I British Airborne Corps was based in Moor Park Golf Club in Hertfordshire. The HQ was never intended to be deployable and hence had not been established as such. There were not enough signallers and those there were had not been on exercise.[115] Once on the ground the role of the HQ was ambiguous. It was usual practice to place an airborne force committed to operations under the command of a suitable ground formation HQ. In the case of Operation MARKET GARDEN this was XXX Corps and Second British Army. The intrusion of Browning's HQ was an extraneous distraction.[116]

Despite these factors Browning landed with his HQ at Groesbeek near Nijmegen on the first day of the operation, 17 September 1944. Predictably, his signals organisation was inadequate and throughout the battle he had very little communication with 1st British Airborne Division, twelve miles to the north. Therefore he could do little to assist and advise Urquhart. By flying to Holland Browning had dislocated himself from the air forces that supported him from England with which he also had only very limited contact. When the battle at Arnhem began to go adversely he was unable to influence the air support plan and adjust the locations of re-supply DZs. Without good communications Browning could do little to influence the three airborne divisions, each fighting its own battle, or the overall air operation being planned and launched from England. In getting himself into that position he had used thirty-eight valuable Horsas to fly in his HQ. 2 Battalion The South Staffordshire Regiment was forced to fly into Arnhem on 17 September 1944 with less than half the battalion because there was a deficiency of thirty-four gliders from the fifty-six required to lift an airlanding battalion. The remainder had to be flown in on the second day of the operation.[117]

Browning's Nijmegen expedition was a poorly judged endeavour that left a blemish on his otherwise worthy airborne career. Shortly after Arnhem, he left the airborne establishment to become COS to Mountbatten in South-East Asia Command. However, true to form, he was soon pushing for the use of airborne forces in that theatre. Down wrote to Gale shortly after Browning's arrival, 'The Boy [Browning] has made his presence felt already. He has covered more ground in three weeks than his predecessor did in the last six months!'[118] He wrote to CIGS on Christmas Day 1944 to explain that 44th Indian Airborne Division, despite his previous efforts, was not yet ready for operations and that there were insufficient transport aircraft available. He argued for the transfer of at least one allied airborne division from Europe to the Far East. He over-optimistically estimated that a fully functioning airborne division might shorten the war with Japan by six months or possibly even a year.[119] 6th Airborne Division was ordered to re-deploy to the Far East in May 1945. The requirement was subsequently reduced to a brigade and 5 Parachute Brigade flew to India in July 1945. I British Airborne Corps HQ, now commanded by Gale, followed a month later to take over planning. However, two weeks later the Japanese capitulated and the Corps HQ was disbanded that October.[120]

Frost was not convinced by the methods used to select airborne commanders during the war. 'It never seemed to me that much serious thought was given to producing the type of leaders or their staff needed to lead a completely new force who would be pioneering a new means of introducing soldiers into battle.'[121] In truth there was little experience or evidence on which to base any selection criteria or a syllabus for training airborne leaders. As with so much of British airborne forces development it followed a process of trial and error. In view of this the competence of the commanders that were produced was very good, particularly up to brigade level where very little evidence of poor leadership exists. At the divisional level and above there are examples of commanders whose decisions and style of command could be questioned. Hopkinson, Urquhart and Browning have been criticised but generally only in respect of isolated decisions and not because of a general incompetence. Hopkinson's decision, against advice, to commit his division on Sicily, and Browning's to take his HQ to Nijmegen were driven by misguided personal ambition as opposed to any gross misunderstanding of the situation. If 1st Airborne Division had not landed on Sicily then many of the faults apparent during that operation might not have been rectified in time for D-Day. Browning's lack of conceptual vision was balanced by his ability to enforce progress and was compensated by Gale's acute understanding of doctrinal development and, more importantly the training issues linked to it, while working as Director of Air in the War Office. Urquhart's poor appreciation of the situation on the first evening of the battle at Arnhem and his failure to apply the tactical principles of airborne warfare, while it had far reaching consequences was an isolated decision made by a commander under intense pressure and with limited experience of the type of battle he was fighting, and at a time where the concept of 'mission command' was largely alien to the British Army. It is also worth considering that while these commanders have to retain the responsibility for the decisions they made many of them would have been informed by the advice they received from their staff.

The Airborne Staff

> Command is not exercised and control is not maintained just simply by the direct issue of orders to subordinate commanders. Between the commander and his troops is a machine, a human machine, the staff.[122]

By the outbreak of the Second World War Gale's words had been fact for many years. The mechanised battle was too diverse, complex and fast moving for a single commander to absorb all the information he required, digest it, make a decision and inform his troops of his intent. Even a battalion commander had a small staff to assist him, each being an expert in his field. The adjutant looked after manning and discipline, the quartermaster had control of the administrative and logistic support and he would have an operations officer and possibly a signals officer and an intelligence officer. Each would routinely gather information from his own area of interest and present it to his commanding officer in a manner that would assist his decision-making process.

At battalion level these officers were selected from amongst the other regimental officers, usually by seniority. At brigade level and above there existed a core of specifically selected and trained staff officers. These were men who had passed the coveted, almost mystical, selection process for Staff College. Attendance at Staff College, preferably at Camberley but also possible at Quetta in India and Haifa in Palestine during the war, was a prerequisite for certain advancement beyond the rank of major. Staff College was not universally esteemed but the criticism was usually groundless and the process successfully selected not only competent staff officers but also those who would go on to the higher echelons of command.[123] Those who attended undoubtedly benefited from the experience and often paid high tribute to the standard of instruction, which was successfully maintained throughout the war years.[124] The British had never maintained a General Staff Corps on the continental pattern. Instead it produced a cohort of able men who were able to alternate between command in the front line and service with the staff in various HQs and therefore gain and spread valuable experience. The cohort, bound by the suffix 'psc' (passed staff college) had a shared experience and often knew each other, and thus the wheels of the Army were often greased by personal relationships.[125]

Within the brigade and higher HQs the staff were essential in order to prevent a commander becoming immersed in detail. Montgomery, who took the system of separating a commander from the trivia to its extreme, had no doubt as to the staff officer's function. 'Details are their province. No commander whose daily life is spent in the consideration of details, and who has not time for quiet thought and reflection, can make a sound plan of battle on a high level or conduct large-scale operations efficiently.'[126] Unfortunately, as Montgomery suggests, the life of the staff officer was normally far less glamorous than that of his counterpart commanding soldiers in a company or squadron. Consequently and unsurprisingly first hand accounts from staff officers, apart from those in the highest positions, are rare. Memoirs from members of the airborne staff are practically non-existent. Both Major A.A.K. Pope and Crookenden actively sought postings onto the airborne staff. Following Staff College a posting to the staff was inevitable and many officers would have made every effort to ensure that this did not keep them out of action. A posting to an airborne headquarters would have fulfilled this ambition. Crookenden went to the HQ of 6th Airlanding Brigade straight from Staff College in mid-1943 and Pope was posted to Browning's staff at the Airborne Divisional HQ in October 1942. Both write about their airborne training, the personalities around them and the operations in which they took part but neither took much time to record the daily routine activity of their HQ or the staff.[127]

Prior to September 1941 the only staff dedicated to airborne forces were those based at Ringway, but those men were employed on administrative and experimental duties as opposed to command and control. During that month 1 Parachute Brigade was formed under the command of Gale and his staff began to gather. With the establishment of 1 Airlanding Brigade the following month a divisional HQ was required to maintain control over both formations. The HQ Airborne Division was formed under Browning during November 1941 with an initial establishment of seven staff officers, which was subsequently expanded to twelve plus an administrative staff the following month. The War Office recognised that in the future the division might be ordered to deploy as a complete formation and therefore the HQ should also be operational and would be brought up to war strength gradually through the ensuing months.[128] By the end of 1941 a provisional establishment for the HQ had been published with the addition of general and intelligence staff and staff officers from the engineers, signals, medical services, ordnance and the chaplain's department. In total there was provision for fifteen officers and seventy-five other ranks.[129]

It quickly became apparent that the duties of Browning and his staff were far more extensive than the relatively straightforward command of two brigades. The Airborne Division HQ became a sorting office for all matters concerned with airborne forces and their development. Constant and direct liaison was necessary with the War Office, the Air Ministry, AC Comd and 70 Group RAF. Browning also had to work with the Airborne Forces Establishment (AFE) at Ringway and with DCO on matters of operational planning. Through his position as the senior British airborne officer Browning also became responsible for Polish, French, Dutch, Belgian and Norwegian airborne contingents beginning to form in the United Kingdom. Before the establishment of Hardwick Hall the HQ was also, administratively if not physically, the *de facto* airborne forces' depot.

Through the first half of 1942 these multiple duties began to exert pressure on Browning and his staff.[130]

Browning recognised the situation and began to lobby HQ Home Forces and the War Office in June 1942 to try to relieve some of the workload from his staff. He recommended that an HQ separate and superior to the divisional HQ should be created in order to take on the liaison and administrative tasks, leaving him free to command his formation. The following month the staff at HQ Home Forces requested that Browning investigate a number of options including the formation of a superior HQ Airborne Forces, but also the establishment of a small separate liaison and administrative HQ to be under control of the division, and also the expansion of the divisional HQ in order to cope with the extra duties it had acquired. Browning reconfirmed his preference for a separate superior HQ but Home Forces dragged its heels and continued to pose questions rather than make a decision, despite his almost weekly correspondence on the subject.[131] In October 1942 Browning submitted a detailed paper to Home Forces precisely delineating the responsibilities between an airborne divisional commander and the desired superior airborne HQ. The duties of the new superior commander and his staff would include liaison with the experimental establishment concerning all trials and equipment development, liaison with allied airborne contingents, liaison with the War Office and the Air Ministry, supervision of an airborne forces depot, the coordination of important visits and perhaps most crucially the provision of an operational airborne advisor. For all this he proposed a HQ under a major general with approximately thirty-five staff, including RAF officers, split into four staff branches of advice and plans, staff duties, training and liaison and administration and quartermaster. This would relieve much of the pressure from the divisional HQ and leave it responsible only for the command and training of the divisional units, liaison with No. 38 Wing RAF for training and operations and liaison with DCO for operational planning.[132]

The experience of 1 Parachute Brigade in North Africa supported Browning's views but it was not until the formation of 6th Airborne Division that HQ Home Forces acceded to the formation of HQ Major General Airborne Forces in April 1943. Browning moved into this new role the following month, handing over command of 1st Airborne Division to Hopkinson.[133] This new arrangement caused some confusion. Chatterton complained that he was 'becoming somewhat mystified about the set-up of Airborne Divisional HQ. General Browning seemed to have been changed from Divisional Commander and had become known as Major-General Airborne Forces.'[134] Indeed it was not long before Browning himself became dissatisfied with the new organisation that he had struggled so hard to establish. On 20 August 1943 he wrote to the War Office complaining that he had no command function and as such, with the same rank as the divisional commanders he was now in a position where he felt subordinate to them. The invasion of Sicily had necessitated his deployment, along with his staff, as airborne advisor to Allied Forces HQ, leaving no means of control back in the UK. He felt his HQ was over-committed and should be enlarged and given the ability and the authority to command more than one division in operations and prepare them during training. Browning recommended a revision of the existing organisation under a Lieutenant General Airborne Forces. On 26 December 1943 Major General Airborne Forces was disbanded on the formation of HQ Airborne Troops, which subsequently became HQ I British Airborne Corps in April 1944. The new organisation was to be commanded by a Lieutenant General and Browning was promoted into the job.[135] By the close of 1944 HQ I British Airborne Corps had a staff of forty-three officers and 203 other ranks.[136] Certainly the establishment of a three star airborne headquarters was justified once there were two divisions in Britain but it cannot be ruled out that Browning had at least one eye on self-advancement when he put forward the changes. Certainly some senior American officers believed that Browning was a scheming empire builder.[137]

During that period of Browning's machinations the airborne divisions were allowed to develop their own HQ and staffs. 1st and 6th Airborne Division's HQ evolved along similar lines through training, trials and the experience gained in the Mediterranean. By August 1943 the HQ of 6th Airborne Division had grown to thirty-two officers and 149 other ranks. This was practically double the establishment of HQ 1st Airborne Division at the end of 1941. A parachute brigade HQ comprised of ten officers and twenty-five other ranks and an airlanding brigade fifteen officers and sixty-eight other ranks. This meant within the division approximately sixty-seven officers and 270 other ranks were involved in the command and control of the formation. The case for such expansion must have been clear to have justified such an investment in manpower. Sixty-seven officers would have filled the establishment of two parachute battalions.

During the Second World War there was a lack of detailed knowledge within the higher echelons of command of the British Army concerning the capabilities and employment of airborne forces. The airborne staff therefore, besides the command and control of airborne formations, had a further role in providing advice and expertise to superior headquarters that were planning, commanding or coordinating the use of airborne forces within a wider battle. This had been recognised as early as March 1941 when the CLE prepared instructions for HQ III Corps regarding the use of parachute troops. Along with useful information on the size and weight of parachutes, containers and trolleys and a checklist of points that should be included in the orders of any operation employing parachute troops it

Command and Control

also identified the requirement for an airborne advisor to be present in the force HQ and gave clear instructions as to his role.

> He must be prepared to advise the Force Commander whether the operation is feasible or not depending on the weather conditions at the time. He should be able to give the proportion of casualties likely due to adverse weather conditions. The Force Commander should then decide on the importance of his task vis-a-vis the probable landing casualties.[138]

At this time only very minor airborne operations could be mounted but as complexity grew so the vital role and necessity of having carefully selected and positioned advisors increased.

In September 1942, 1 Parachute Brigade was designated for employment during Operation TORCH, the allied invasion of North Africa. The Brigade was allotted to First British Army under General Sir Kenneth Anderson and the Brigade Commander, Brigadier Edward Flavell, dealt directly with HQ First Army. Browning and the divisional HQ assisted but had to concentrate on a myriad of tasks to prepare the Brigade for war. Browning repeatedly requested that a senior airborne advisor should be appointed to HQ, First Army but this was not granted. The result was that during both planning and execution of the operation in North Africa Flavell was pulled between commanding his brigade and assisting with planning at HQ First Army. Inevitably this led to neither appointment being fulfilled satisfactorily as he could not be in two places at once. The Combined Chiefs of Staff learned this lesson, stating that it would never work well to employ senior officers from airborne or troop carrier tactical units to act temporarily in a higher staff capacity as doing so would 'seriously cripple the combat efficiency of the tactical unit concerned'.[139] The lack of a permanent airborne expert in the superior HQ led to 1 Parachute Brigade not being employed at full capacity or in the most suitable manner. There were also glaring omissions in administration such as a lack of maps and aerial photographs and in the coordination between the British paratroops and the American troop carriers. All these points were detailed in the post-operational report, including a recommendation to increase the staff in a brigade HQ so as to cope with ground operations and simultaneous preparations for airborne operations.[140]

Many of the lessons were learned prior to Operation HUSKY, the allied invasion of Sicily in July 1943. The position of Major General Airborne Forces had only just been agreed to in principle by the War Office when Browning was appointed Airborne Forces Adviser to HQ 15 Army Group. He was responsible for advising the higher commanders and coordinating inter-services requirements with the RAF and Troop Carrier Command. 1st Airborne Division was placed under command of Montgomery's Eighth

Army. Once given his mission and tasks it was Commander 1st Airborne Division, Hopkinson's duty to give advice to Montgomery and cooperate directly with his staff for planning and cooperation. Meanwhile Browning remained responsible for matters such as provision of stores and equipment, aircraft routing, photographic cover and offensive air support.[141] For these tasks he had a staff of just four officers. This division of responsibilities, while created with the best of intentions of relieving HQ 1st Airborne Division of some its peripheral workload, allowed dangerous gaps to appear in the staffing process and certainly assisted Hopkinson in his machinations to deploy his Division whatever the cost. When Browning left the theatre at the end of July he attempted to close these gaps. He recommended the establishment of a permanent Mediterranean airborne staff. By combining staffs in North Africa and the Middle East he believed a pool of fifteen officers covering all disciplines could be created. Crucially the task of this staff was to embed itself in various HQ at all levels, including Eighth Army and XIII Corps. This staff would report up to a Brigadier General Staff (BGS) who would work between AFHQ and HQ 15 Army Group. This organisation ensured closer coordination and meant that a commander could not plan an operation without some degree of scrutiny from a superior HQ.[142]

If the 'outstanding tactical lesson' of Operation HUSKY was the potentiality of airborne forces then the major lessons identified by the airborne establishment were those of coordination and staff effort. Browning concluded that much of the staff work concerned with an airborne operation must be centralised at the highest appropriate level so as to allow coordination between the land and air components under a single commander.[143] The AFHQ report supported the view of its senior airborne advisor.[144] The COS Committee also considered the airborne contribution to the operation and emphasised the vital importance of joint planning and particularly the role of the air force. Airborne operations should be planned integrally within the overall air operation both strategically and tactically. It had to be controlled tactically by the same formation controlling other air operations within the immediate vicinity. The controlling influence of the air force during the preparation, planning and sanctioning of operations was reiterated. In order to assist in the planning and execution of individual operations airborne advisors should be appointed to the air force commander's staff.[145] Browning's analysis of the structural requirements of the future airborne staff were essentially sound, based on a joint controlling headquarters at the highest level and airborne advisors embedded where necessary in all subordinate headquarters.

It became clear from Operations TORCH and HUSKY that any future major airborne operations were going to be allied affairs requiring the close integration of British and American airborne and air forces. The Combined Chiefs of Staff recognised this and stated that there must be a commander

Command and Control 167

with a staff who was able to directly control all troops and aircraft involved in an allied airborne operation. They were critical of senior commanders, claiming that they had generally failed to follow established principles, and re-emphasised that the planning of airborne operations could not be successfully delegated without losing coordination. They identified this as being primarily the result of 'higher commanders [that] have failed to provide themselves with capable airborne and troop carrier staff advisers'.[146] The result of this edict was the creation of the First Allied Airborne Army, initially outlined in early June 1944 and formally established that August under Lieutenant General Lewis Brereton, an American Air Force commander. Brereton's HQ controlled five, later six, British and American airborne divisions, and crucially, IX US Troop Carrier Command and 38 and 46 Groups RAF. The functions of the HQ were laid down by Eisenhower and included the training, development, re-supply and reconstitution of airborne formations, the preparation and examination of outline plans for the employment of airborne forces and the direction and control of such plans until a designated ground force commander took over. Not everyone was in favour of the new organisation. Leigh-Mallory thought that increases in personnel and thus time and labour would inevitably lead to a loss of efficiency. He believed that the reorganisation was illogical and unsound on the basis that the previous method of operating had not failed. He was also unhappy that the air force elements of the new establishment would not be available for other purposes between airborne operations.

Despite this opposition HQ First Allied Airborne Army was established but not in time to influence the planning for D-Day, which followed the same process as that used prior to Operation HUSKY. The outline airborne plan was developed at SHAEF and passed to the army group HQ for development. It was then disseminated to the relevant army HQs and on down to corps, including the airborne corps HQ. Therefore it was not until the orders permeated down to army and corps HQs that joint planning was conducted with airborne and ground force commanders together. At this level there was no control over the air forces involved and therefore any changes to the air plan would have to be staffed back up the chain of command. This system was cumbersome but worked successfully for the invasion of Normandy due to the protracted planning period. With the establishment of HQ First Allied Airborne Army joint planning could take place from the very outset, with advisors available direct to SHAEF. Brereton's HQ would then be placed under the command of a particular army group for specified periods. Thus all its resources were allotted to 21 Army Group for the pursuit from Normandy and during the Rhine Crossing. With Brereton embedded at this level initial outline plans for the employment of airborne forces could be scrutinised from the ground and air force perspectives simultaneously. Thus First Allied Airborne Army HQ acted as a 'planning filter', checking the validity of proposed airborne plans and then passing them down to the airborne corps HQ with much of the coordination already in place.[147]

HQ First Allied Airborne Army, with its constituent XVIII US and 1 British Airborne Corps, developed a successful planning and coordination mechanism that was effective throughout the remainder of the war. Valuable lessons had been learned and huge advances had been made since Flavell had struggled to both command his brigade and act as a staff officer at HQ First Army. The internal brigade and divisional staffs had also evolved into robust and practical organisations in a relatively short space of time. This situation was, however, not unique to airborne forces. Conventional infantry and armoured formations also had to rapidly increase and develop their staff after inter-war neglect. For example in 1936 the GOC of 1st Division in Aldershot had only three staff officers permanently attached to his headquarters. Even this was better than his territorial contemporaries who had only two.[148] Whereas in the case of a commander it could be argued that long service with airborne forces was an advantage, with the staff a breadth and variety of experience was often a benefit. Obviously detailed airborne knowledge was essential while planning, but having a wider range of backgrounds and cap badges within the headquarters often assisted the coordination with ground forces. To that end airborne headquarters benefited from the British staff system, as did the rest of the Army. The German Army system of maintaining a specialist staff created a gap between 'those who thought and those who fought', while the British shared and moved around an individual's experience.[149] Take Crookenden as an example. He was a pre-war, regular infantry officer with eight years' experience before he attended Staff College. He then took that experience to HQ 6 Airlanding Brigade during the planning, training and execution of the invasion of Normandy. After Normandy he was promoted and moved to command 9 Battalion, The Parachute Regiment where he would have been able to draw from his knowledge of how a brigade HQ thought, planned and operated.

Individual Commanders and their Staff as a factor in Development
Britain's new airborne capability required innovative methods of command and control in two distinct key areas. At the higher level a joint approach to command, staff and headquarters was needed so that the effective coordination between the necessary land and air forces was in place, thereby ensuring efficient planning and close control of operations. Ideally a single commander should have the necessary land and air forces under his direct and sole control. This was not achieved until August 1944 with the establishment of Brereton's HQ First Allied Airborne Army. It was not mature enough to effectively influence the plan for MARKET GARDEN as demonstrated by the lack of coordination during the selection

of DZs and LZs. However by March 1945 and Operation VARSITY it could plan, execute and control the highly efficient airborne element of the Rhine crossing all within fifteen days.[150] The burden of the detailed coordination required to achieve this was carried by the airborne staff. It may appear that the airborne staff had a less stressful existence than their counterparts in armoured and infantry formation headquarters, thanks to the low incidence of airborne operations throughout the war. However the number of cancelled operations also has to be taken into account. Around fifteen operations in France and Holland were planned and cancelled between June and September 1944. Thirty-two airborne operations were initiated and cancelled by Alexander's headquarters in Italy between March and May 1944. At least five projected airborne operations in Germany were cancelled between March and May 1945. Despite not being launched the planning for each of these operations required no less staff effort, while seeing that effort continually failing to reach fruition must have had an adverse affect on morale. This unglamorous burden was borne stoically by officers who in peacetime might have been successful barristers, solicitors and schoolmasters.[151]

Before the creation of First Allied Airborne Army coordination with the air force had to be conducted by land commanders at the operational level in order to be effective. However most of these commanders had little practical experience of airborne operations and had to rely on advisors and be prepared to study the problems and characteristics inherent in airborne warfare. Where an operational level commander failed to apply any conceptual thought to the employment of airborne forces their use became more difficult and less probable, Alexander in Italy being the obvious example. Montgomery and Eisenhower did both apply time and effort to understanding the nature of airborne warfare and while they were not entirely coincident on the subject the principles they produced were valid and enduring. These principles were important to the development of Britain's airborne forces at the operational level. Where they were ignored the odds of tactical success were reduced, as at Arnhem, but more significantly those principles had the potential to influence the wider airborne concept of employment, as will be examined in the following chapter. As Liddell Hart observed, successful training, organisation and planning for operations demanded, above all, intellect based upon knowledge.[152] However, at the most senior levels it is difficult to justify criticism of commanders for displaying inadequate personal knowledge of the employment of airborne forces. This was a novel concept and most senior officers during the period between Dunkirk and D-Day would have had more pressing matters to attend to than the study of an arm which they might or might not at some point in the future have under their command for a brief period. When it became clear to a commander that he would have control of airborne forces for a particular operation then there is clear evidence that time and thought was applied, as was the case with Dempsey, Crocker and O'Connor during the build up to D-Day.

The second characteristic of airborne warfare that required an innovative approach to command was apparent at the tactical level. The maroon beret and the parachute badge were physical representations of the 'imperative of example', a signal that the tactical airborne commander had endured the same risks and hardships as his men during training. However, training soldiers to react according to their own initiative could make command by traditional methods difficult. A looser framework of command was required, an ethos already adopted by the Germans as *Auftragstaktik* and recognised currently within the British Army as 'mission command'. Although the limited published airborne doctrine did to some extent recognise the unique tactical circumstances that an airborne force might have to fight under it did not go so far as suggesting any distinctive style of command necessary to cope with them. Notwithstanding the lack of any formal direction there is certainly ample evidence of a style of leadership close to what would now be termed 'mission command' being adopted by airborne commanders while the majority of the British Army 'retained its commitment to autocratic, top-down managerial control'.[153] That evolution of command must therefore have been driven from the bottom up as a reaction to the environment that airborne commanders found themselves in rather than through adherence to any official doctrine. This perhaps explains why this looser style of command, while often in evidence at the lowest tactical levels, occasionally appeared not to extend to the higher tactical level around divisional and corps command. This is unsurprising when it is considered that it would be decades before the British Army formally recognised the tenets of mission command. However, Gale felt that the failure of the British to adopt a style of command more akin to the Germans during the Second World War was at least partly to blame for many of the tactical setbacks experienced by the British Second Army in Normandy.[154]

The 'imperative of example' was a double-edged weapon. Officers made up less than 6 per cent of an airborne division's manpower.[155] This figure is approximately equal to that found across the infantry during the Second World War. However in 6th Airborne Division in Normandy officers made up over 9 per cent of the men killed in action and 11 per cent during the Rhine Crossing. 1st Airborne Division officers contributed 16 per cent of those killed during the battle at Arnhem.[156] These figures are not necessarily unique to airborne formations (Brigadier Lord Lovat's 1 Special Service Brigade lost 53 per cent of its officers during eighty-three days of fighting in Normandy) but the rate of loss was greater than for most other ground formations.[157]

These casualty rates were practically unsustainable. Only one in three brigade commanders and one in nine battalion commanders re-crossed the

Rhine with 1st Airborne Division during the initial evacuation from Arnhem. It was this scale of loss within such a short time frame coupled with the rapid expansion of the airborne establishment that necessitated non-airborne officers being posted into positions of formation command. There is no evidence that these commanders performed any better or worse in general than their 'thoroughbred' counterparts, although two in particular are open to criticism.

The critical comments of Hopkinson and Urquhart made in this chapter are obviously the product of hindsight. They are not a general reproach of either of the officers' abilities but represent a focused assessment of their actions at specific points in time. These observations have to be tempered by factors influencing the two men at the time. Nevertheless, Hopkinson's blinkered insistence on committing his Division to the invasion of Sicily and Urquhart's failure to drive 1 Parachute Brigade forward on the first evening of Operation MARKET GARDEN were significant incidents. They both directly influenced the development of tactical doctrine as applied to subsequent operations, hence the scrutiny is justified.

The same criteria apply to the criticism of Browning. Clearly his decision to deploy HQ I British Airborne Corps to Nijmegen was poorly judged at best. His deployment prevented a battalion from flying to Arnhem complete and more significantly deprived the operation of an informed controller in a position to influence the battle. His lack of conceptual insight could have had a detrimental effect on airborne development if there had not been an officer of Gale's calibre in a position to pick up the reins. His insistence on discipline could be viewed as contrary to the tenets of 'mission command' but it paid dividends during the 1st Airborne Division's withdrawal from Arnhem and 6th Airborne Division's protracted defensive battle in the Orne bridgehead. He immersed himself in detail that might have overwhelmed others but he forced progress and produced results. Browning's performance can be assessed far more favourably if he is measured as a senior staff officer and administrator rather than as a commander. After all, his positions as the first GOC 1st Airborne Division, Major General Airborne Forces and Commander I British Airborne Corps were essentially created to improve coordination and development, not to lead troops into battle. Above all Browning set himself up as a figurehead for Britain's developing airborne forces. To the men below him he became 'the distillation' of the airborne establishment and inspired such loyalty that many sprang to his defence following his unflattering portrayal in the film 'A Bridge Too Far', twelve years after his death.[158]

CHAPTER SIX

Concept and Doctrine

Conceptual and Doctrinal Innovation

The physical and moral components of fighting power are underpinned by a third, conceptual component concerning principles and understanding: the concept and doctrine of a fighting force. This aspect of war fighting is vital as it provides the common framework within which different units, formations, arms, services and even nations can conduct operations with a degree of coherence and mutual understanding. When an innovative military capability or technology was introduced, as many were during the war, it took time for this mutual understanding to develop. An appreciation of the potential and limitations of the capability could diverge between different areas of the military establishment, based on their own prejudices and wider interests.

The terms 'concept and doctrine' do not necessarily translate directly into the language understood by the military establishment during the Second World War. A concept is a description of the way in which a military capability will be employed within a given environment. It describes the function or purpose of that capability in a manner that allows its development to be framed and parameters set for the procurement of equipment and training of personnel. It is a high level document that then has to be 'translated into a doctrine that will provide sufficient guidance for the force to use in its war preparation without being so specific that it binds too tightly the hands of the future commanders who will have to use it'.[1] A concept informs developers while doctrine guides practitioners. A concept prescribes where and when a capability will fight while doctrine advises how it should fight. The two together therefore are the catalyst through which physical resources under command are translated into military effectiveness.

Although the terminology might have been unfamiliar during the first half of the twentieth century the idea of framing a concept ahead of drafting doctrine was commonplace. Some conceptual debates have attracted a good deal of scrutiny, such as the purpose and function of the tank within mechanised and armoured warfare from 1917 through to 1943.[2] To produce an innovative concept requires intellectual as well as professional vision. Individuals with vision and the position, aspiration and ability to communicate that vision and drive the organisational change required to implement it are a rare commodity.[3] Churchill was such an individual but his initial minute in June 1940 failed to adequately express his vision. It fell far short of articulating an adequate concept of the purpose that he expected his proposed airborne force to operate within. As development progressed Churchill was never able to dedicate sufficient intellectual space to drafting any sort of concept to explicate his original vision.

Without the Prime Minister's contribution it was the task of the officers of the War Office, the Air Ministry and the central staff to develop the airborne concept. Recent studies have concluded that the British Army was far more open to innovation than it has previously been given credit for.[4] However, this was most apparent at the micro or tactical level. Within the higher echelons of the establishment the Army and, to a lesser extent, the RAF remained inherently conservative and strictly hierarchical, inducing a lack of imagination, excessive caution, professional pessimism and conventional thinking.[5] It has been concluded that, 'Those being ordered to innovate may well not have control over everything needed to carry out the order, particularly if what is needed is unconventional creativity.'[6] With the collective military intellect focussed on the immediate threat to national survival during the early part of the war 'unconventional creativity' was not a luxury that the over-burdened staff could afford.

The purpose and function of the new airborne force had of necessity to be centrally dictated to ensure that the concept integrated with and contributed to overarching war plans. Doctrine however required a degree of critical examination. It was designed to guide the preparation and actions of practitioners but it relied on the intellectual honesty of those practitioners to provide reaction and comment based on the experience of exercises and operations. Nevertheless, at some point the debate has to halt and an enduring core to tactical doctrine has to be identified, which is retained despite adaptations and amendments following lessons gathered from its application.

Developing the Airborne Concept
Limited thought had been applied to the potential of airborne warfare in and around the British military establishment before the Second World War. In an early example of 'unconventional creativity' Lieutenant Colonel Groves of the Royal Flying Corps foresaw in 1917 an end to the 'battering ram tactics' of trench warfare. He believed that 'a new and tremendous phase in air warfare is about to begin – the phase of the Long Arm'.[7] Groves described his 'Long Arm' concept in detail: how a force of 15,000 men

might be landed from the air in the enemy's rear area simultaneous with an attack on his front line. This force would then destroy enemy batteries and disrupt his reinforcements and supplies. Groves also explained how an airborne force might be used to strategic effect by landing a force well behind the enemy front, close to industrial areas in order to disrupt factories and manufacturing. Fear of such action would cause an enemy to deploy troops to guard such installations thus entailing attrition of his reserves and therefore of his front line.[8]

Nearly two decades later official British military observers attended Soviet demonstrations of its parachute troops.[9] The War Office attempted to assess the potential of this new capability during a brief flurry of staff activity that followed the Soviet airborne exercises in 1935. Just over a year after the Red Army demonstration near Kiev, the initial British attempt to express a theoretical concept of employment for airborne forces was not encouraging. The General Staff identified three different types of operations, which might conceivably be carried out by a force landing from the air:

(a) The landing of a comparatively large force with the intention of maintaining it by air.
(b) The landing of a force, whether it is large or small, without any intention of maintaining it by air, but in the expectation that the development of operations on land will shortly permit its maintenance by land.
(c) The landing of a small force in the nature of a 'forlorn hope', which will do as much damage as possible and subsequently surrender.[10]

Whilst the use of the term 'forlorn hope' is seldom reassuring, this list does not describe the function of a future airborne force. It more closely resembles an early attempt to write a doctrinal statement, a foundation for the practical application of an airborne force. The function of that airborne force in the wider sense remains unclear. Less than eighteen months later the General Staff did produce a more clearly defined conceptual statement. It concisely identified the activity to be accomplished by an airborne force.

(a) The seizure of important points in rear of the enemy lines, in conjunction with an advance or attack of land forces.
(b) Surprise action in the pursuit.... The [airborne] group may be expected to hold its ground until relieved by the advancing forces.
(c) Attacks on enemy headquarters, centres of communications, e.g. railway junctions or important bridges, L. of C. [lines of communication] installations, aerodromes etc.[11]

This marked an improvement; however, these ideas had been formulated for exercise purposes only and amounted to a projection of how British airborne forces, which had not as yet been conceived, might in theory operate.

While the War Office was first attempting to grapple with the concept, Liddell Hart also contributed to the subject. In 1936 he wrote of the physical and psychological strain resulting from having to defend rear areas and vital points from an airborne assault.[12] A year later he included in a list of important technological military advances, 'The development of parachute forces which may be dropped ahead of an advancing army to secure key points, or may be dropped in the enemy's rear to seize points on his communications.'[13] Liddell Hart would contribute to the debate again but in 1937 his thoughts, like those of the War Office, were academic. British airborne forces would not be conceived for another three years and in the meantime there were more pressing and basic military requirements to be determined. Hence in 1940, immediately following Churchill's initial minute, a mature concept of the function of British airborne forces, if an aspiration at all, was still distant.

In September 1940 the Joint Planning Staff (JPS) began to study possible future offensive operations. They developed five scenarios (four of which had been suggested by Churchill) all of which involved seizing and developing a bridgehead on an enemy held coast. The five operations outlined were the invasion of Norway, the Low Countries, France, the Iberian Peninsula and metropolitan Italy, the latter of which was considered the priority.[14] By 26 September 1940 the JPS had produced the first draft of their paper 'Future Plans: Basic Requirements', the remit of which was to detail 'recommendations regarding certain basic requirements which are likely to be common to all operations [involving seizing a bridgehead]'. The JPS suggested that the early capture of aerodromes on a hostile shore was essential, both in order to deny their use to the enemy and in order to maintain them for allied use. The planners declared that the capture of enemy aerodromes was a suitable task for airborne troops.[15] Aerodrome capture groups thus became listed as an integral part of any future invasion corps.

The JPS certainly held both the position and the authority required to devise a concept and therefore influence the manner in which Britain's fledgling airborne forces should be utilised. However, 'Future Plans: Basic Requirements' detailed the capabilities required to achieve a large-scale invasion of mainland Europe across all arms and services. Airborne forces were not accorded any special priority by the Joint Planning Committee (JPC). Additionally there is no evidence to suggest that anyone in the JPS in 1940 had the intellectual vision or 'unconventional creativity' to foresee the full future potential of the airborne capability. The JPS had to adopt a more pragmatic methodology and to rely on experience to guide their concept for airborne forces. In September 1940 the only experience available on which to base their plans was that of Germany. On 9 April

1940 German airborne forces seized airfields at Aalborg in Denmark and at Stavanger and Oslo-Fornebu in Norway. On 9 May 1940 the War Cabinet sought assurance from the Dutch government that their airfields had been prepared for demolition in order to prevent them from being captured intact. However the next day German airborne troops seized aerodromes at Waalhoven and Katwijk in The Netherlands.[16] German airborne forces were part of the *Luftwaffe* rather than the army and therefore these airfield assaults represented discrete 'air operations for an air purpose'.[17] Control and integration were relatively simple due to the single chain of command for the air and ground elements of the operation. As the Germans discovered, using airborne troops to support land operations such as those that seized the crossings over the Albert Canal presented a more complicated set of doctrinal challenges in order to achieve the necessary integration and coordination. To the JPS airborne forces represented an expedient solution to the problem of aerodrome capture, providing them with a discrete capability that would neatly accomplish one of the objectives that must be achieved as part of a successful invasion.

The degree to which the British military establishment relied on German experience to frame its own concept can be clearly traced in early doctrinal publications. 'Military Training Pamphlet No.50, Airborne Troops, Part I – Defence Against Airborne Troops' was published in 1941.[18] The sixteen-page document detailed the characteristics, organisation, tactics and equipment of German airborne troops. It also outlined the counter measures required to defend against an airborne operation.[19] A comprehensive section dealt with the defence of airfields against airborne attack, which the pamphlet regarded as likely employment for German paratroops. The types of operation on which it was considered German airborne forces might be employed were broken down into major and minor roles. The definition of these roles was a word for word copy of those listed for British airborne forces in 'Army Training Instruction No.5, Employment of Parachute Troops', also published in 1941.[20] The implication was that British doctrine would be an exact reflection of German doctrine. This was a tenuous assumption considering the dearth of information on which it was based. The British military establishment had little firm evidence of the doctrine and tactics used by German airborne troops in the Low Countries and Norway in 1940. The British Army did not face German paratroops and airlanding troops until the invasion of Crete in May 1941. A comprehensive report was produced on German airborne military effectiveness and doctrine displayed on Crete but it was published too late to influence the writing of either MTP No.50 or ATI No. 5.[21]

On 5 October 1940 the initial draft of 'Future Plans: Basic Requirements' was approved by Churchill.[22] The paper became the accepted overarching concept for Britain's future offensive operational aspirations, and airborne forces formed an integral part of that concept. By the time the full document was published two weeks later the JPS considered that 'unless a proportion of air-borne troops are included in the proposed Invasion Corps, the difficulties of certain operations we are considering might be considerably increased'.[23] However, aerodrome capture was not universally accepted as a suitable or credible purpose for airborne troops. Even before the JPS had published their paper the concept had been questioned. Sceptics in the Air Ministry doubted whether the German experience could be repeated and considered that there was very little prospect of replicating the successful airborne operations conducted by the Germans in 1940, particularly if it was intended to capture an aerodrome as an essential precursor to further air operations.[24] The analysis behind this pessimism was logical and difficult to deny. The German airborne operations in 1940 represented technological surprise at the operational level. It was unlikely that the Germans would allow themselves to be surprised in a similar manner and their airfields on the continent would probably be heavily defended as a consequence.[25] Not only was future aerodrome capture considered to be an extremely high risk concept it was also deemed by some to be unnecessary. The Air Ministry was beginning to recognise the potential use of gliders and concluded that being able to land heavy equipment almost anywhere on the battlefield by this method reduced the necessity to seize airfields in the initial stages of an invasion.[26]

Despite some doubts as to its validity, 'Future Plans: Basic Requirements' continued to contain the aerodrome capture concept when it was published and controversy persisted for at least another eighteen months. The General Staff were indecisive from the start. In January 1941 they appeared content that the most likely purpose for airborne troops lay in the capture of aerodromes.[27] Only days later they had changed their mind: 'As regards the purpose for which airborne forces are required, it is thought that they should not be primarily related to the capture of aerodromes but that the primary reason should be to enable the land forces to occupy enemy territory.'[28] One member of the Army was in no doubt that the airborne concept was a far more complicated and potentially more significant capability than had been examined thus far. On 24 June 1940 Major J.F. Rock, a Royal Engineer, commissioned in 1925 and a graduate of Staff College, was commanded by the War Office to take control of the development and organisation of Britain's emerging airborne forces. It is difficult to ascertain why Rock was singled out for this job as nothing in his previous record suggests that he might be regarded as an expert or even particularly air minded.[29] Despite a lack of experience Rock does appear to have had a degree of intellectual vision when it came to recognising the potential of the airborne capability, in particular gliders and airlanding troops.

Rock was certain that aerodrome capture was only one of the many tasks for which an airborne force was suitable and probably the one that was least credible. During recent command conferences at Camberley senior officers had discussed how the Germans might employ airborne forces during an invasion. Aerodrome capture was not one of the tasks that they considered likely.[30] He suggested a broader concept of purpose including attacking defended positions from the rear in conjunction with a conventional frontal attack, isolating the enemy from reinforcement, seizing important bridges and defiles, flank attack, a feint to draw off enemy reserves and the disruption of communications. Before the end of the war, British airborne forces would conduct most of these types of operation. If Rock had a degree of vision he certainly had the time to commit to developing the airborne concept: as the senior Army officer at the CLE it was his only remit. However, what Rock lacked was the position to influence development at the higher level. He could develop tactics but as a junior major and then lieutenant colonel, isolated at Ringway hundreds of miles from the War Office, he could do little to influence the overarching concept. The JPS did not involve Rock in the planning process and hence the work carried out by his Tactical Development Section at Ringway was not heeded as it 'had not been related to a specific operational object'.[31] On 8 October 1942 Rock died following a glider accident and the British airborne establishment lost one of its more prescient thinkers.

Rock realised that using previous German experience on which to base their likely future intentions during an invasion of Britain was not credible. The Air Ministry had also come to the conclusion that aerodrome capture as a purpose for airborne forces had been based on the unsound extrapolation of enemy experience into allied future operations. Goddard believed that 'the demand for the creation of airborne forces has arisen from our experience of what the enemy has done with them.... Perhaps a too ready assumption follows that our requirements are similar. In the exercises which have taken place I doubt whether a sufficiently realistic British point of view has prevailed.'[32] The Air Staff went further, suggesting that the entire basis on which the requirement for aerodrome capture groups had been based was now invalid and that the Invasion Corps project was already out of date.[33] By the spring of 1941 the War Office and Air Ministry were agreed that they had learned false lessons from the use of German airborne troops in the Low Countries and that it was doubtful whether such ideal conditions would recur.[34] However, despite it being pointed out that enemy air strength and ground defences would make the capture of aerodromes by lightly armed airborne troops a high-risk endeavour, the task persisted in published doctrine but was placed further down the list of priorities. The isolation of the enemy from his reserves and the attack in the enemy rear in conjunction with a conventional frontal attack, as proposed by Rock, now appeared as more probable concepts of purpose.

Concept and Doctrine

It was German action that would signal the end of aerodrome capture as a valid concept, just as it had created it in the first place. On 20 May 1941 the Germans attacked Crete, initially using an almost exclusively airborne force. In the morning gliders and paratroops landed at Maleme and the airfield there was assaulted. Despite having a high degree of air superiority it took the German airborne units thirty-six hours to capture the airfield and even then it was not fully secure. A British intelligence report outlined the German experience.

> On the whole from the German point of view the results obtained from the employment of parachutists in Crete must have been below expectations. At Canea, Retimo and Heraklion, the parachutists did not achieve the rapid success expected and their losses were out of all proportion to any local advantage gained. It must be acknowledged, however, that the capture of Maleme landing ground, which was the turning point in the campaign, was effected largely by parachutists. Even here, however, the situation remained critical for two days and was only held by intensive air support given by an airforce which enjoyed, for the time being, complete air superiority. Without this, the attack would have failed.[35]

Had the defenders been more adeptly handled Maleme airfield might have been denied to the Germans. It was a costly victory with 4,000 Germans killed or missing and 175 of the 530 transport aircraft employed destroyed or damaged beyond immediate repair.[36] The lessons were clear; the Germans had attempted to repeat their experiences of 1940 without fully realising that the operating environment had changed. It appeared that the Germans were no longer sure about the correct method of use of their airborne forces and made several fundamental errors. It was foolhardy to drop paratroops onto heavily defended airfields. They suffered heavy losses and would have been far more effective had they been dropped on the island out of range of defending fire.[37] Gale was certain that the risk involved in capturing an airfield, even with air superiority and in the face of incoherent defence, was now close to unacceptable.

> The almost complete lack of fighter defence or adequate anti-aircraft artillery gave to this operation [Crete] a sense of unreality.... In Crete reinforcements [sic] of the defence was impracticable, whereas in northwest Europe the German counter-attack could be both massive and rapid and airborne assault troops might in these circumstances be liquidated before relief could arrive.[38]

Following Crete, aerodrome capture began to fade as a valid purpose for British airborne forces. Lord Cherwell, Churchill's chief scientific advisor, believed that if aerodrome capture persisted as a task it would require the development of new weapons to ensure success and even then it might fail in the face of a determined counter attack.[39] Where aerodrome capture did continue to appear as a concept it was in a modified form as in 'Airborne Operations Pamphlet No.1 – General' published in 1943.[40] In late 1942 it was considered a realistic task for airborne forces to assist in achieving air superiority by denying to the enemy the use of forward fighter aerodromes.[41] Landing on an airfield and denying it, through cratering for example, is a far simpler operation than seizing it intact and then defending it with light weapons.

In retrospect it is initially difficult to understand, except for the sake of expediency, why the JPS considered aerodrome capture not only as a credible purpose but as the only conceptual purpose for airborne forces during an invasion. The concept, based on operations in an environment that was unlikely to recur, was clearly flawed. However, officially at least, it was the only concept considered for many months and was still retained, although as a lower priority, after the Germans had demonstrated the high risks involved. The adoption of and perseverance with a concept and purpose based on false lessons from previous operations is indicative of the lack of clear and coherent conceptual thought concerning British airborne forces in the War Office during the early part of the war.

Aerodrome capture is not the only example of poor conceptual expression associated with Britain's airborne forces. The new airborne capability was clearly going to have most utility in the offensive role. However there were two distinct areas of offensive planning being undertaken in the early part of the war. First there were operations designed to strike the enemy in his rear areas, destroy strategically important targets, harass and cause the enemy to re-deploy reserves to cover the threat. These types of operation were likely to be transient in nature but could be mounted early in the war effort and might be termed 'deep' operations in today's parlance.[42] The second class of operation was a major offensive involving the mass invasion of mainland Europe and the 'close' engagement of the enemy on his front line. These operations would probably be enduring but could not hope to executed until the requisite trained manpower and equipment had been amassed. There was no explicit distinction made between the two types of operation except in terms of where command and control was invested for planning and execution.

Airborne forces obviously had utility in both 'deep' and 'close' operations. Groves in 1917 and the War Office in 1936 and 1937 both including references to types of 'deep' and 'close' operation that could be conducted by an airborne force. However, influencing the 'close' or tactical battle by isolating an enemy from his reserves or attacking him in the rear is clearly a different type of operation to attaining 'deep' or strategic objectives by destroying, neutralising or isolating vital military, civil or political targets. If both were to be adopted as valid purposes for airborne forces then each would require its own doctrine in order to ensure that both could be executed effectively and efficiently. However the differences between the two were at best poorly expressed and at worst not even recognised until the latter part of the war. That lack of distinction was in turn responsible for imprecise doctrine, which manifested itself frequently throughout the war.

Early attempts to express on paper future objectives that might be achieved by airborne forces in both 'deep' and 'close' operations ranged from the confused and misguided to suggestions bordering on the ridiculous, as this list from October 1940 demonstrated.

(i) The immobilizing of large numbers of enemy troops in dispositions unfavourable to their strategy.
(ii) The spearhead for offensive action within a range of say 500 miles of a suitable air base.
(iii) As a self contained force, capable of being sustained by airborne supplies, for small localised actions.
(iv) For the 'planting' of agents, saboteurs and other irregular troops in enemy territory.
(v) For air transport by towed gliders of personnel, rations and equipment thus greatly augmenting the scope of usefulness of operational bomber aircraft.
(vi) The bringing into the War effort to a greater extent the woodworking trade (furniture etc) which could undertake the building of gliders.[43]

Although a product of the CLE, this is not the work of Rock. The first statement above was a sound expression of the effect on the enemy resulting from the threat of airborne forces operating in the deep concept. The second statement, though valid in itself, did not progress the close concept, as it did not include an objective or purpose for the 'spearhead'.[44] Statement three once again contained no objective or purpose and could have been taken to mean either deep or close employment, although the former was more likely to have been the intention because of reference to the action being small and localised. The fourth statement was, in effect one means of achieving the objective expressed in the first. It confused legitimate employment within the deep concept with tasks that would come outside of the remit of airborne forces. This was probably the result of the ambiguous language in Churchill's original minute. It is difficult to give the fifth statement any credibility as part of an airborne concept, particularly as the Air Staff were adamant that airborne forces would reduce the operational effectiveness of the bomber force. This was an example of the misguided interpretations of the employment of gliders, which will be expanded later in this chapter. Finally, statement six bordered on the ridiculous. The warlike employment of the furniture industry might be a useful by-product of the development of airborne forces but it could not realistically have been considered one of the latter's functions. Reintroducing the Trojan Horse as a weapon of war would have had a similar effect.

Blurring the line between 'deep' and 'close' employment is explicable when the lack of experience, knowledge and vision concerned with the subject in 1940 is considered. However, in at least one example it was the result of a deliberate policy. The Air Ministry believed that there was no need to distinguish between airborne forces destined to act as a spearhead for an invading force, those required for tactical operations in conjunction with a land battle and those employed on sabotage and espionage operations.[45] This might have been true when considering the recruitment and individual training of airborne forces; however it was not a sound principle to apply when developing concepts, doctrine, equipment and collective training.

If the conceptual ideas coming from official channels were confused then there were men on the periphery of the official military establishment who were producing innovative and coherent thinking. Just before the outbreak of war Liddell Hart believed that the greatest potential for airborne employment would be in Asia and Africa where physical communications were poor. In Western Europe he concluded that an airborne force would probably be quickly located and overwhelmed by motorised troops and tanks and therefore their use in that theatre was unlikely.[46] In 1941 however, following the German example, Liddell Hart contributed with a far more lucid and tangible vision of the airborne concept.

> If the primary aim of the invading force is to secure some key city or port, it would be essential for the invaders to gain control of the main roads leading in that direction before the strategic defiles on the way – the river crossings or mountain passes – can be occupied by the defenders reserves.
>
> There are risks to air-borne detachments being landed so far forward. But they are outweighed by the risks which the whole invading force, and its mission, would run, if the defender were allowed time to occupy these barrier-points in any considerable force.
>
> On this calculation, the invading commander may well decide to use the bulk of his parachute forces for an 'attack in depth', and a lesser part for attacking the rear of the beach defences.
>
> On this calculation, the invader may conclude that the key to success for him lies in *jumping into the interval* (between coastal defences and mobile reserves), with his parachute troops, and blocking the move-up of counter-attack divisions, before they can get properly underway.[47]

Concept and Doctrine 183

This was a perceptive conclusion from Liddell Hart. The importance of judging and then controlling 'the interval' in terms of both geography and time became apparent in Sicily, in Normandy and during the Rhine Crossing. However, by the time this was written he had lost the sponsorship of the Secretary of State for War and his ideas and comments on the subject were too brief and infrequent to influence those responsible for airborne development. Although the isolation of reserves did appear as published doctrine in 1941 the intended objective of the root concept, as described by Liddell Hart above, did not.

Another protagonist in the debate was Leo Amery. He accurately drew the boundary between the 'close' and the 'deep' concepts in late 1941 while trying to gain support for Indian airborne forces. He attempted to explain the differences in employment in North-West Europe as compared to the Indian sub-continent.

> The Germans used them [airborne forces] at Rotterdam and in Crete, for tactical purposes, i.e. for intervening in the actual battle.... I should have thought that, on the other hand, the greatest value of airborne troops, at any rate in an area like the Middle East, lay in their strategical [sic] use, i.e. in being able to send off appreciable forces to seize distant positions in advance of our own movements, or behind the enemy's rear, where no serious opposition to landing need be expected.... After all, it was this strategical [sic] use of airborne troops, on a very small scale indeed, reinforcing Habbaniya, landing behind the Iraqis at Falluja, and sending off a detachment to Mosul, that proved so useful in the Iraqi campaign.[48]

This was a remarkably prescient observation. Amery saw the utility of airborne forces in India in 'deep' employment: being dropped on key vulnerable points in order to carry out internal security operations and therefore achieve strategic effect i.e. the enduring stability of colonial rule. In fact by the time Indian airborne forces were ready to be committed they were only employed in the 'close' concept for 'tactical purposes' at Rangoon during Operation DRACULA in May 1945.

In 1943 Otto Miksche, a Czechoslovak officer serving with the Free French Forces on the staff of General de Gaulle in London published 'Paratroops', the first attempt to outline the history, utility and future of the airborne capability and place it in the public domain. By 1943 the 'deep' concept had essentially become the domain of special forces such as the SAS and SOE and Miksche did not consider 'deep' operations in his book, describing only 'The Tactical Employment of Airborne Troops.'[49] Miksche split the 'close' concept of operations into two categories based on their effect on the enemy, direct and indirect, and offered several examples to illustrate this idea including a forced river crossing and an amphibious landing.⁵⁰ In both cases he cited the capture of key defiles in the enemy's immediate rear in order to prevent the movement of enemy reinforcements as indirect action. This was a reasonable concept and corresponded with Liddell Hart's 'interval'. However, as direct action he quoted the immediate seizure of an enemy held riverbank or beach. This might be reasonable if the operations were only lightly or unopposed but to attempt to capture a defended riverbank or beach would rank alongside aerodrome capture in terms of risk.

Despite aberrations such as this, Miksche, along with Liddell Hart and Amery, all demonstrated some degree of intellectual vision when it came to the conceptual employment of airborne forces. However, the thoughts and ideas of these men were never pooled and used to make up for the lack of 'unconventional creativity' in the traditional military establishment.⁵¹

During the early work carried out in the War Office and Air Ministry an implication began to emerge that the two different concepts of operation for airborne forces, 'deep' and 'close', were inextricably linked to the size of the formation required to carry out the task. The implication was made explicit in 1941 when it was stated that airborne operations fell into two different categories.

Major.
(a) Seizure of an advanced tactical position, to be followed up by supporting forces.
(b) A major raid, where the airborne forces will be withdrawn after completion of the task by land, sea or air.
(c) Emergency air transportation of troops etc.

Minor.
(d) Sabotage.
(e) Espionage.
(f) Subversion.
(g) First flight of an amphibious operation.⁵²

Once again the different concepts according to which airborne forces could be employed had been confused by an attempt to draw the boundaries according to the size of the operation or the formation taking part. The first statement fell within the close concept whereas the raid, whether major or minor, as outlined in the second statement would have normally sat more comfortably in the second group of operations. As to the 'first flight of an amphibious operation' being a minor operation, Sicily and Normandy would demonstrate the error behind this classification. All of the preceding naive attempts to express a concept for airborne forces were at least harmless while they were constrained to papers that circulated between the Air Ministry and War Office. In fact the exchange of ideas, however ill conceived, must be considered healthy considering the lack of a single, visionary force behind the development of the airborne concept. However, the result was a confusion of immature principles that could potentially be dangerous if they became published as doctrine and presented as policy to the wider military community, as they were in ATI No.5 in 1941.

Notwithstanding this, the confusion and debate surrounding the differentiation between major and minor concepts were largely academic up until the end of 1942. The airborne concept during that period would be dictated by resource limitations rather than through doctrinal debate. By the time Operation TORCH was launched in November 1942 Britain's combat effective airborne force was represented by a single deployable parachute brigade. The first thirty months of development had therefore produced a force of approximately 8,500 men with light weapons only, less than half of whom could have been considered militarily effective. Earlier during development the limitations were even greater. By June 1941 only 18 officers and 320 paratroopers were immediately available for operations. However the number that could be dropped on any given operation was considerably less at just 56, restricted by there only being seven Whitleys converted and available to drop airborne troops.[53] A force of less than 60 men could not be considered practical for anything other than a minor operation and in 1941 minor equated to deep operations or raids. The manpower and equipment situation improved only slowly and the employment of British airborne forces within the deep concept from 1940 to 1942 was therefore the result of resource constraints rather than a conscious conceptual or doctrinal decision.

The only other option open to the airborne establishment was to withdraw from active operations during this period, concentrate on training, equipment procurement and tactical development and wait for an operational imperative to be created such as a full scale invasion of mainland Europe. Isolating and retaining a military force purely for research and development is difficult to justify while the rest of the establishment is simultaneously employed in prosecuting a major war. Reserving resources in terms of men and equipment purely for experimentation and innovation, even on a relatively small scale, is unlikely to be accepted, particularly by a nation on a total war footing. The non-contribution of the early airborne force to the war effort attracted adverse attention and parts of the military establishment were keen to recoup some of the valuable resources invested in the airborne experiment, particularly in terms of high-quality personnel. As early as August 1940 Dill pointed out that the commandos and paratroops under training consisted of specially picked men and contained a high proportion of officer material. The Army was badly in need of such material and, if there was no immediate prospect of them being used for offensive operations, he wanted to have high quality manpower returned from the commandos to their units.[54] Churchill denied Dill's request on this occasion but if further attempts to recoup valuable manpower and aircraft committed to airborne forces were to be countered then there had to be a visible physical contribution to the war effort. Isolating themselves during a protracted period of experimentation and development was not a viable option.

Although small scale, deep raids became the default concept and function for Britain's airborne forces until the end of 1942 the control mechanism for planning and executing airborne raids during that period was not within the airborne establishment's remit. Following Dunkirk, regardless of the lack of resources, Churchill did not want Britain to settle into a defensive mindset. He charged DCO with turning 'the south coast of England from a bastion of defence into a springboard of attack'.[55] However, in 1940 DCO was essentially a naval organisation, with only amphibious manoeuvre as a means of delivering any necessarily small scale offensive or raid on to enemy territory. Thus combined operations were restricted to the coast and prone to predictability. The potential for British airborne forces to improve the limit of combined operations beyond the littoral environment was recognised by DCO: 'Unless arrangements for airborne raids are ensured, offensive activities against the enemy must necessarily be limited for a very long time to areas lying within a mile or two of his seaboard.'[56] However, DCO contained neither the inherent expertise nor the experience required to plan airborne operations. It had to rely on advisors such as Browning to inform its decisions and planning process. But Browning and other senior members of the airborne establishment were biased and eager to see their arm employed whenever practical in order to promote its capabilities, justify the resources invested and keep the War Office and Air Ministry at bay. After the success of Operation BITING, the Bruneval raid in February 1942, DCO sought every opportunity to deploy airborne forces. During 1942 airborne participation was planned but cancelled for at least five operations, the most potentially damaging of which was a considerable contribution to Operation JUBILEE, the raid on Dieppe.[57] Browning encouraged DCO throughout this period.

> As a final request, I urge most strongly that Airborne Forces may be given far more opportunities of taking part in or initiating raids on the continent. A great deal of experience is required before airborne tactics and administration generally can be improved…. There is no reason known to me why we could not carry out with our own aircraft, within three weeks from date of warning, operations using two companies of parachute or light airlanding troops…. Such operations would be extremely good for morale…. I would like to aim at one small operation

per fortnight, with two or three larger scale ones later on during the winter. Given the organisation and staff asked for, this can be done.'[58]

Not only were Browning's figures over-optimistic, but his approach was potentially detrimental to development. First there would be a constant attrition of trained manpower and equipment that would slow progress and could, in the case of the raids failing, undermine confidence in the entire capability. This factor was understood and stressed by more perceptive members of the airborne establishment. For example, Rock considered that 'both parachute troops themselves, the aircraft which carry them and the aircraft crews are too valuable to be used up on minor operations not directly connected with an offensive against Germany and Italy…. British parachute troops should not be used for anything less important than say the capture of Channel ports, as a preliminary to an invasion of France or in a major offensive against the Italians in North Africa.'[59]

The difference in opinion between Browning and Rock is indicative of the dilemma that faced Britain's airborne establishment. On one hand there was a desire to take part in necessarily small-scale raiding operations in order to demonstrate the utility of airborne forces and justify the resources invested in the programme. On the other was the longer term aspiration to conserve the physical capability and train men and procure equipment to enable the participation in large-scale operations as envisaged in 'Future Plans: Basic Requirements'. This dilemma persisted because there was no controlling authority in a position to resolve it. Without an established joint organisation or individual charged with coordinating airborne development there was no coherent concept of purpose during the period from 1940 to the end of 1942. This was not only the primary factor in the rate of development during this period but was also critical in dictating military effectiveness.

The pernicious effect of the absence of joint direction and control, and therefore a coherent concept for employment, became apparent in the tactics employed by Britain's airborne force during its first contribution to a major operation. TORCH should have been a watershed – the point at which the airborne capability moved from being restricted to minor DCO operations to being a component of a full scale conventional battle. Instead 1 Parachute Brigade's contribution to the operation was characterised by 'fervent though amateurish efforts'.[60] The three separate battalion-level British airborne operations conducted during TORCH were akin to three individual raids. The airborne battalions were dropped far ahead of British First Army's front line, up to three hundred and fifty miles in one case, beyond the point at which they could directly influence the close battle, and were expected to be self sufficient once on the ground. They had no fire support, received no re-supply and evacuation of casualties was practically impossible. The experience available within the Headquarters of First British Army on which to base airborne planning was severely limited and had to be supplemented by moving 1 Parachute Brigade's commander, Brigadier Edward Flavell, onto the staff. Therefore, in the absence of a robust controlling authority tactical doctrine was driven by the airborne troops and their commanders on the ground. The experience of those at the tactical level had been accumulated entirely from planning and conducting deep raiding operations and hence the British airborne performance in North Africa reflected that experience. Britain's airborne forces had adopted a doctrine of 'quickly in and quickly out'.[61] The absence of both detailed intelligence and the means of expeditious extraction, and the presence of a mobile and heavily armed enemy, made 'quickly in and quickly out' all but impossible. Frost's battalion lost two hundred and sixty men killed, wounded or missing during a single five day period of operations.[62]

The Development of Airborne Tactical Doctrine

Strategic decision-making changed the military environment during the allied conference, SYMBOL, held at Casablanca in January 1943. The conference endorsed the Mediterranean offensive policy and designated Sicily as the next Allied objective. The principles and conditions required for a major opposed amphibious assault had been defined by the JPS in 'Future Plans: Basic Requirements' at the end of 1940. Although the scale of HUSKY would be greater than the corps level operation envisaged by the JPS, the fundamental requirements remained extant and these included the utility of an airborne contingent. Thus a large-scale operational imperative for Britain's airborne forces was created and the impetus for development was firmly established. With HUSKY being planned for the summer of 1943 progress became expeditious. Churchill's 'stand still' order was rescinded and barriers to development began to fall away. The requirement for a much larger airborne force was accepted and as 1st Airborne Division trained for HUSKY the War Office issued orders for the formation of 6th Airborne Division in April 1943. The reliance on the volunteer system of recruitment was no longer paramount as the Army recognised the need to man the airborne establishment quickly and earmarked infantry battalions for wholesale conversion. The rate of rise in trained airborne manpower became exponential.

Although SYMBOL did not result in the creation of a joint organisation specifically to monitor and guide airborne development, the operational imperative did begin to align the War Office and Air Ministry's approach to development. From the beginning of 1943 onwards the Air Staff no longer opposed airborne forces on an institutional level, although individual officers did continue privately to doubt their utility.[63] The Air Ministry could no longer resist the operational imperative for airborne forces and acceded to the War Office's requirement for support aircraft. As 38 Wing RAF was expanded to 38 Group the provision of transport aircraft became a more imperative issue and monthly deliveries of the Dakota steadily increased until 46 Group RAF was established in January 1944, equipped solely with that aircraft.

Despite this renewed impetus there were still barriers to overcome before HUSKY could be executed. Notwithstanding the Air Ministry's improved efforts, the provision of suitable RAF aircraft to drop paratroopers and tow gliders from North Africa to Sicily still did not meet the requirements of the operation. As with TORCH Britain's airborne forces employed during HUSKY had to rely heavily on American aircraft. While the solution to this physical problem was relatively straightforward, there was also a doctrinal issue that required the application of deeper consideration. HUSKY was the first operation during which appreciable airlanding forces were employed even though the tactical doctrine for a large-scale glider assault was untested. The only previous use of gliders had been during operation FRESHMAN but the raid on the heavy water plant at Ryukan in Norway had been a disaster, with the all the airborne personnel involved either dying in crashes *en route* or being captured and then executed by the occupying Germans.[64] Even if FRESHMAN had been successful TORCH had demonstrated that basing the tactical doctrine for large-scale operations on the experience from minor raids was disadvantageous. In addition to this lack of experience the tactical employment of gliders and airlanding troops had been the subject of enduring and often specious debate.

The possibilities of gliding for military advantage had stirred official interest long before the War Office briefly examined the potential of parachute troops from 1935 to 1937. As early as 1922 the Air Ministry was taking an interest in gliding at the highest level. Air Chief Marshal Sir Hugh Trenchard as CAS personally directed more effort to be put into investigating the potentialities of the military use of gliders in August of that year.[65] At the end of September 1922 Squadron Leader Maurice Wright and Captain W.H. Sayers were despatched to Fulda in Germany to observe glider trials. Clearly they were not impressed by their observations and transmitted this opinion to their masters in the Air Ministry. Despite some of the gliders taking part achieving impressive feats of endurance in the air, Wright concluded that there could be no direct military or commercial value attached to gliding.[66] With growing public interest in the subject, thanks to a £1,000 prize being offered by the Daily Mail in a gliding competition, the Air Ministry felt compelled to make a statement. Published in the press, the official account of the Fulda trials concluded that prolonged flight by gliders over definite distances was not possible and this fact precluded them from becoming a dependable method of transport.[67] A proposed glider committee was abandoned and Trenchard and the RAF lost interest in the subject.[68]

At the outbreak of the war there was no British military gliding activity whatsoever. This remained the case until mid-1940 and the German invasion of the Low Countries. Following this event the possibility that gliders might be used during an invasion of Britain began to be considered. A small experimental Special Duty Flight was established at Christchurch to ascertain the probability of gliders being detected by radar. This small organisation became the basis of the initial glider training establishment following its move to Ringway.[69] However, basing tactical doctrine on the German model for glider operations was an even more tenuous proposition than it had been with paratroops. There were doubts about whether gliders had been used at all during the German operations in Belgium and The Netherlands. A month after the event the Air Ministry intelligence department had only unconfirmed and unreliable information that gliders had been used during the capture of the Belgian fort at Eben Emael. Stressing that this was only speculation, the Air Staff conceded that it had no knowledge whatsoever that any real military value was placed on the glider by Germany.[70] Three months later the Air Ministry was more certain that gliders had been used during the capture of 'a defended position in Belgium', but there were no further details of the methods or tactics used.[71] Even as late as August 1942 the role of the glider during the Eben Emael operation was unclear. An American publication related that only paratroops had been used and the German demolition teams had arrived by road after the initial *coup de main*.[72]

With no obvious source of 'unconventional creativity' and no clear German example to copy, the War Office and the Air Ministry struggled to express a coherent application for the glider as a military tool. They realised it must have a useful application, otherwise why would the Germans have used them, but they followed blind alleys and fanciful postulations rather than first adopting the obvious and simplest solution. Suggestions were put forward that gliders could be used to drop very large bombs of two or three tons.[73] Using gliders as towed fuel tanks to enable air-to-air refuelling of long range bombers was also proposed.[74] However, amongst the unlikely and unfeasible schemes the true value of the glider was eventually identified. The carriage and delivery of light tanks and other heavy equipment was suggested as a possible suitable role for investigation. It followed that if gliders could carry heavy equipment then it was reasonable to suppose they could be used for landing troops. In September 1940 the Air Staff brought some sense to the proceedings. 'We must walk before we can run. At the moment we have no experience in operating towed gliders…. I would hesitate to put out requirements for gliders for these purposes [tank lift, refuelling etc] until we have obtained some experience with gliders for the transport of troops.'[75]

Although the more outlandish proposals were quickly dispensed with, the true capability of these aircraft took longer to determine. All aspects

Concept and Doctrine   191

of the glider's technical potential were a mystery in mid-1940 and some misapprehension remained into 1942. For example it was believed in June 1940 that a glider could not be towed off the ground without some form of assisted take-off, and well into 1941 there were still discussions regarding the length of runway required to get airborne.[76] In June 1940 it was believed that 'fairly good' weather was required to operate gliders with no cloud below 10,000 feet.[77] As late as October 1942 it was still considered that the presence of cloud made the use of gliders impossible.[78] However the most potentially damaging assertions in June 1940 were that gliders could not be used in large numbers together and could not land at night.[79] This view may have been an aberration as elsewhere in the Air Ministry the opposite analysis had been made.[80] Nevertheless the same points were still being put forward for clarification twenty-eight months later.[81]

Of course these questions could be effectively answered by running trials and a series were ordered at the end of 1941 to look at mass landings and landings in confined areas and at night.[82] However, sometimes the true capability of the glider was discovered by accident. The forced landing of a Horsa in April 1942 was written up as an example of just how effectively the aircraft could land in a confined area if necessary.[83] The problem with conducting a sufficient number of trials frequently enough to expedite the fact-finding process was the availability of aircraft, particularly the gliders themselves. The Horsa and Hotspur were not available for trials until the summer of 1941. In the spring of 1941 frustration and concern over the lack of progress with glider development was beginning to become apparent in the War Office and the Air Ministry.[84] The lack of firm evidence demonstrating the glider's true potential and limitations allowed enthusiasts and dissenters to voice their opinions without their facing the possibility of being contradicted by facts. The arguments put forward during this period, primarily by the Air Staff, were based purely on theory and projection and did little to advance progress.

Another feature of these debates was that the glider was often placed in direct competition with paratroops, with participants extolling the virtues or constraints of the one versus the other, as if the two were mutually exclusive. In the summer of 1940 the Air Staff were already beginning to realise the scale of the task of providing aircraft for paratroops. Perhaps as a distraction from these difficulties the Air Ministry suggested gliders as an alternative rather than a complementary capability to paratroops. With very little verified information and no practical experience on which to base it, the Air Staff's assertion appears to have been a reckless assumption.

> We are beginning to incline to the view that dropping troops from the air by parachute is a clumsy and obsolescent method and that there are

far more important possibilities in gliders. The Germans made excellent use of their parachute troops in the Low Countries by exploiting surprise, and by virtue of fact that they had practically no opposition. But it seems to us at least possible that this may be the last time parachute troops are used on a serious scale in major operations.[85]

The Air Ministry then vacillated between support for either gliders or paratroops for another two years, and in September 1942 the doubts expressed over the utility of gliders, supported by Cherwell and Cripps, were largely responsible for causing Churchill's decision to drastically reduce airborne development. Much of this conniving was due to the Air Ministry endeavouring to preserve the means to conduct its own core doctrine. The promotion of gliders ahead of paratroops and the subsequent complete shift of opinion was an attempt by the Air Staff to reduce the number of bombers it would have to commit to the developing airborne forces. However, the fact that it often used doctrinal theories and assumptions to advance this aim impeded the efforts of those who were earnestly trying to develop airlanding tactics and policy. Many of the Air Staff's arguments rested on an adversarial approach to the integration of the glider and the parachute; it was either one or the other. This spread into wider doctrinal thinking. Often when airborne doctrine or tactics were under discussion the potential of airlanding was weighed against that of parachuting, rather than evaluating the two side-by-side as complementary capabilities. The advantages and disadvantages of paratroops were listed beside those of glider-borne troops without considering how the advantages of one might offset the disadvantages of the other.[86] The 'major' and 'minor' concepts further divided them. It was supposed at one point that parachutists were more suited to 'minor' operations while gliders should invariably be used for large-scale, 'major' operations.[87] The German experience appeared to support completely the opposite view, with gliders being suited for the capture of specifically designated and locally defended objectives while parachutists were seen as more effective for the purpose of capturing large areas.[88]

Efforts were made to separate parachutes and gliders by time and space on the battlefield. Most theories along these lines proposed an initial landing by paratroops subsequently reinforced by a glider landing.[89] This followed the accepted, although incorrect, perception of German experience. The official history relates that the Germans 'always realized that troops carried in aeroplanes or gliders had a great advantage over parachute troops.... However, it was accepted that normally parachute troops must land first in order to secure landing places.'[90] Although on Crete gliders were used in the first wave of the assault, attempting to achieve surprise by casting off from their tugs at a distance from the coast. The sequence of the attack was gliders then parachutists, followed by air transported and then amphibious troops.⁹¹ In fact preceding a glider landing with paratroops was sound doctrine. Parachutists could ensure that there was no substantial local opposition, remove obstructions, give the signal for the main air landing operation to take place and lay out wind indicators and boundary marks.⁹² This eventually became the role of specially trained pathfinder units within airborne formations. However, early attempts at integrating gliders with paratroops were based on the supposition that this would be the only role for the latter. This theory was expounded in considerable detail by some proponents and led to the assumption that airlanding troops would always greatly outnumber their parachute-borne compatriots in any airborne operation. The proposed proportion of glider troops to parachutists varied from three to one up to ten to one in favour of airlanding troops.⁹³ The actual proportion within an airborne division later in the war would be just over one to two in favour of paratroops.⁹⁴

Despite these machinations and the lack of experience, the plan for the employment of airlanding troops during HUSKY was both credible and sensible.⁹⁵ 1 Airlanding Brigade was deployed by glider on the night of 9/10 July 1943 to capture the Ponte Grande over the Anopo Canal by *coup de main* and then secure the adjacent town of Syracuse. Gliders were also used to fly in heavy equipment to support 1 Parachute Brigade's assault on the Primasole Bridge over the River Simeto four days later. However, despite this relatively mature approach to the tactical employment of airlanding troops and the success of the operations the cost was excessive, particularly during 1 Airlanding Brigade's operation. In some cases as few as 34 per cent of the gliders landed on their designated LZs, with over 50 per cent ditching in the Mediterranean.⁹⁶ This in turn led to the high casualty rate in the Airlanding Brigade of nearly 500 men killed, drowned, wounded or missing. In total the Division suffered over 700 casualties during the operation.⁹⁷ The unacceptably high casualty rate among British airborne troops on Operation HUSKY was due almost entirely to the inexperience of American aircrew and a lack of collective training immediately prior to the operation. However, the board of inquiry instigated by the AFHQ, with limited airborne experience and distorted by multi-national and inter-service sensitivities, came to different conclusions. Despite both Browning and Alexander identifying the lack of collective training as the key factor, the official report blamed the use of small tactical DZs and LZs and their proximity to the enemy objectives for the high casualties.⁹⁸ The result of this flawed analysis, despite the success of the airborne operations on Sicily, was a shift away from accepting risk to aircraft and men during the early stages of an airborne operation.

The conclusions of the report became absorbed as implicit doctrine within the airborne establishment. Although there is no evidence that changes to tactical doctrine were formally directed the tactical expression of this implicit doctrine was apparent during the invasion of Normandy. With the exception of the *coup de main* on the Orne and Caen Canal bridges, 6th Airborne Division's DZs and LZs on D-Day were large, often brigade-sized pieces of ground. Most were at least a mile and many up to three miles from the units' objectives. This factor combined with the scattered night drops reduced the margin of success in some areas to a very narrow degree. Without the cover of night the reduction in speed and surprise caused by the distances involved could have been critical. However, the overall achievements in Normandy masked any tactical weaknesses and the level of scrutiny and analysis conducted after the operation was low.

Headquarters First Allied Airborne Army was activated in August 1944 but there is no evidence that it conducted any focused post-operational analysis of OVERLORD, and in any case it was launched immediately into a frantic planning cycle attempting to keep pace with the advance of the allies across Europe. With no further analysis the apparent success in Normandy appeared to validate the implicit doctrine adopted after HUSKY.

During the planning for Operation MARKET GARDEN the continuing shortage of aircraft and concerns over losses from flak led to the pattern of large DZs and LZs selected well away from enemy influence being stretched and repeated in order to further reduce the risk at the front end of the operation. In Normandy, one to three miles between landing and the objective severely reduced the margin of success, despite the protracted planning period and overwhelming fire support. Such luxuries were not apparent at Arnhem where the distances involved were extended to six to eight miles. The margin for error had been reduced to such an extent that when the 'partially known' factors, such as weather and the enemy and 'the unknown factors commonly described as luck', went against Urquhart the chances of success became very slim indeed.[99]

As well as Urquhart's personal leadership style a further doctrinal factor also exacerbated the flaws in the plan at Arnhem. During training, airlanding and parachute battalions and brigades were separated at unit and formation level. This was a historical legacy that had administrative and training advantages. The parachute and airlanding capabilities had evolved separately. 1 Parachute Brigade was formed in September 1941, before any airlanding units had been established, and hence it retained its integrity as a parachute only formation. 1 Airlanding Brigade was converted wholesale from 31 Independent Brigade Group and hence there was a valid reason for not immediately breaking up that formation while it adapted to its new role. Keeping the two roles separated by formation suited their respective requirements for individual training. The equipment, organisation, real estate and time required to train paratroops were very different from those of an airlanding unit and the essential glider pilots. This all made eminent administrative and operational sense up to the point of landing. However, once on the ground a parachute unit or formation brought quite different capabilities and qualities to the battle than its airlanding counterpart. A parachute battalion had three fighting companies totalling 556 men whereas an airlanding battalion had four companies totalling 864 men. A parachute brigade had integral engineer and medical support. An airlanding brigade had anti-tank batteries and a light anti-air battery as part of its organic support.[100] Parachute units were light and trained to move quickly in the assault. Airlanding units had heavier firepower but with it came an inevitably enlarged logistic tail that could slow movement.

This in itself was still not a great problem. Even today different capabilities are still generally separated on unit or formation lines during peacetime in order to ease the administrative and individual and specialist training burden. However, for collective training and operations, units and sub-units are moved and exchanged in order to ensure that a formation fights with a balance of capabilities. This is a process known as battle-grouping. Battle-grouping as a concept was not completely alien to the British Army during the Second World War. For example, although the low level doctrine and tactics may have taken some time to perfect, the advantages of a squadron of tanks being attached to an infantry battalion for an attack or vice versa had been widely recognised since operational experience gained in North Africa, although official doctrine took some time to reflect it.[101] However, battle-grouping, making the most of the advantages of both parachute and airlanding troops, was seldom part of the airborne planning process. A notable exception occurred on D-Day when a reinforced company of 2 Battalion The Oxfordshire and Buckinghamshire Light Infantry was attached to 7 Parachute Battalion to seize the crossings over the Orne River and the Caen Canal. However this was more a consequence of the capability of the glider, which was able to land a compact group of soldiers close to the targets silently, rather than of the men inside them.[102]

The consequences of a lack of planned battle-grouping became pronounced at Arnhem. 1 Parachute Brigade and 1 Airlanding Brigade were both dropped on the first day of the operation on separate but adjacent DZs and LZs. 1 Parachute Brigade's task was to move east and capture the bridge or bridges in Arnhem. 1 Airlanding Brigade's task was to defend the DZs and LZs several miles west of Arnhem until the second drop the following day. This meant that while ten heavily armed airlanding companies were allotted a static, subsidiary task, only nine lightly armed parachute companies were committed to the Division's main effort, seizing the bridges.[103] Admittedly the parachute units were more likely to generate the speed in the assault that was essential to reach the bridges but they were landed fifty minutes after the Airlanding Brigade. It is worth considering what the effect might have been of attaching one airlanding company to each of the parachute battalions, thereby putting twelve companies on the main effort and still leaving seven, totalling approximately 1,500 men, to defend the DZs and LZs for twenty-four hours. These airlanding companies could have been ready to join 1 Parachute Brigade by the time it was landed and able to begin the advance east. As the parachute battalions forged east at best speed they could have identified areas of enemy resistance and called forward an airlanding company to deal with them while the paratroops bypassed the enemy and continued their advance. This is, needless to say, a retrospective appreciation but one that would appear relatively intuitive to a commander today. Battle-grouping did occur later in the battle but it was through force of circumstance, rather than by appreciation and planning, as the integrity of many units was lost through numbers of casualties and the Division fell back into defence.

The result of both the flaws in the operational planning and the tactical doctrine, despite Montgomery's assertion that MARKET GARDEN was 90 per cent successful, was an unmitigated disaster at Arnhem which led to an entire airborne division being written off and unfit for operations for eight months. It also exposed the flaws in the implicit doctrine adopted after HUSKY. Unsurprisingly the desperate outcome of 1st Airborne Division's battle at Arnhem was the cause of a good deal of introspection. In retrospect Urquhart recognised that the balance of risk had swung too far in favour of preserving men and in particular aircraft during the early stages of an operation. 'It would appear a reasonable risk to have landed the Div[ision] much closer to the objective chosen, even in the face of enemy flak.... An extra two minutes flying time in the face of flak, if not too severe, would have put the Div[ision]... much nearer its objective. Initial surprise in this operation was obtained, but the effect of the surprise was lost owing to the time lag... before the troops could arrive at the objective chosen.'[104]

Now, with Headquarters First Allied Airborne Army firmly established and a relatively quiet period in terms of airborne planning through the winter of 1944, the lessons from Arnhem could be thoroughly analysed from a joint perspective. Airborne warfare had to return to the doctrine of Sicily – small tactical DZs and LZs close to the objectives – and in doing so the reservations and caution of the air forces might have to be overruled, shifting the balance of risk back to the front end of an operation in order to increase the overall chance of success.

Once again these lessons became implicit doctrine to be applied during the next airborne operation. The contribution of 6th Airborne Division to Operation VARSITY in March 1945 was a model of airborne warfare. The Division landed in a single lift, deploying all three brigades and the organic support within one hour. Many of the troops landed practically on top of their objectives in the face of flak and fire from the enemy on the ground.

The paratroops and airlanding troops in most cases landed in small, tactical company-sized groups with separate designated objectives, supported by heavier weapons landing by glider in the same area immediately after them. As a result all the Division's key objectives had been seized within five hours of the initial landing. The link-up with the ground forces, which had begun crossing the Rhine in advance of the airborne operation, was achieved within the same time span. In addition another doctrinal lesson, apparent to German airborne forces for some time, had also been assimilated. During the Rhine crossing the airlanding troops were used to secure point targets, the bridges over the Issel River, while the parachute formations were used to occupy areas of ground to disrupt the movement of German reserves. Despite the relatively high casualties (nearly 350 killed and over 630 wounded from 6th Airborne Division on the first day) the advantages of landing in tactical groups as close to the objective as possible was stressed. The post-operational report endorsed the principle of accepting more risk early. The use of airlanding troops for pinpoint, *coup de main* type operations and parachute troops to occupy larger enemy held areas was also confirmed.[105] More significantly the critical contribution of a controlling joint headquarters in efficient operational planning and effective post-operational analysis had been demonstrated and validated. Finally, by March 1945 this effective system resulted in credible and potentially enduring tactical doctrine.

The Airborne Concept, Tactical Doctrine and Military Effectiveness
The concept for the employment of Britain's airborne forces and the tactical doctrine that guided their operations are the line of development that had the most obvious and direct impact on airborne military effectiveness. However, as has been shown the concept and doctrine adopted at any point during the war was a result of the influence of factors stemming from the other lines of development. From June 1940 until December 1942 the concept in which airborne forces were employed was that of minor, deep raiding operations. However, the adoption of this concept was through expediency and necessity rather than through any deep routed principles or because of a long-term plan for the development of the airborne establishment. The lack of resources in terms of the persistent dearth of support and transport aircraft and to a lesser extent the initially slow accrual of trained manpower limited the size of a force that could be employed.

In turn both the deficiency of aircraft and the pressure on manpower that forced the need to take part in raiding operations were largely political factors. Throughout development Britain's airborne forces had no single person or organisation who was able to influence progress at a political level. Churchill was too busy to give the project more than intermittent interest. Renwick's terms of reference were too narrow to allow him to exert control in all the areas of development that required coordination and oversight. The Chiefs of Staff's views on the subject were irreconcilably divergent to ever allow progress at the rate that was required. Without an individual empowered to drive the change, which was unpalatable to the ministries and departments in a position to expedite progress, the Air Ministry, MAP and to a lesser extent the War Office were able to pursue their own agendas in relation to airborne development and afford it a commensurately low priority. Hence the provision of the physical resources required to progress airborne development continued to be a detrimental factor.

In addition to the difficulties in coordinating the provision of equipment and manpower the lack of a clearly defined control mechanism also made conceptual development problematic. A concept had to be enduring in order to provide a firm platform upon which the developmental process could be built. However, the airborne concept until December 1942 wavered between the aspiration to build towards large-scale 'close' operations and the necessity to execute minor raids. Conceptual red herrings such as aerodrome capture were allowed to persist. With no clear, centrally generated concept practitioners had to conceive and evolve their own ideas. This could be effective and Montgomery and Eisenhower's concepts of mass and deliberation proved to have enduring credibility.

SYMBOL established the operational imperative for expeditious developmental progress from January 1943 until February 1945, which had not been instituted by any nominated individual or organisation prior to that period. However, resources continued to influence development. The provision of aircraft never recovered to the point where the RAF was able to deploy a British airborne division into battle in a single lift. To overcome this one of two approaches was possible for operational planners. First was to rely on allied, principally American airpower. This became a feature of all major British airborne operations from TORCH through to VARSITY, with the exception of those conducted on D-Day, where the RAF was able to effectively fly shuttles across the Channel. The use of American aircraft during HUSKY, the unfamiliarity of their crews with British airborne techniques and the lack of time for collective training to remedy this directly contributed to the high casualty rate. The second approach was to stagger the fly-in of an airborne formation. This was the case during operation MARKET GARDEN and this factor along with the reluctance to accept risk to the aircraft during the deployment seriously reduced the probability of tactical success being achieved.

This pernicious and continued influence of limited aircraft availability on military effectiveness was exacerbated by there being no centrally coordinated and focused scheme for operational analysis following each operation. The was no organisation such as the AORG in airborne terms to gather information, conduct post-operational analysis and then use the results to influence continued development. This is manifest in the lack of primary evidence demonstrating centrally coordinated improvements to equipment and training as a result of lessons learned during operations. Instead observations through experience were translated into changes from the bottom up in a haphazard manner. The development of the leg bag for carrying personal equipment is an example of this. Similarly, changes to tactical doctrine were adopted implicitly following each operation rather than through any process of explicit revision. The risk with implicit doctrine is that it may not be common across a force and it may even differ between formations. Doctrine is open to individual interpretation and encourages false assumptions about others' understanding.[106] Notwithstanding this risk, if implicit doctrine is published and distributed locally it can induce a degree of mutual understanding and common practice in a given formation or theatre. One example was a Standard Operating Procedure (SOP) produced by Supreme Headquarters, Allied Expeditionary Force (SHAEF) on 13 March 1944 and amended on 8 June and 4 November the same year. Memorandum No.12, 'SOP for Airborne and Troop Carrier Units', was published to provide 'a common basis upon which the training and operations of allied airborne and troop carrier units can be conducted'.[107] The document detailed liaison requirements, staff and operating procedures and the responsibilities (including joint responsibilities) of airborne and troop carrier commanders. The SOP was clearly amended on the basis of experience gained in Normandy and during Operation MARKET GARDEN and its final amendment was made once First Allied Airborne Army had been established.

However, the fact remains that the implicit doctrine adopted following HUSKY led, in part, to the disaster at Arnhem. This could perhaps have been averted had an organisation been established at a level that could command and control both the air and airborne forces, oversee all airborne operations and thus provide a structured approach to doctrinal development. HQ First Allied Airborne Army was such an organisation but its creation came too late to influence military effectiveness until VARSITY in February 1945.

Gale summarised the conduct of British airborne operations during the second half of the war: 'Except that all were airborne, no two operations [between 1943 and 1945] had been alike. There was, thus, no set-piece method of employment for this new arm.'[108] This is an uncharacteristic lack of insight from Gale because clearly there were enduring methods of employment at both the conceptual and doctrinal level. Mass and deliberation were enduring and speed and surprise were invariably doctrinal requirements at the tactical level. What is true is that these features were not adequately identified, emphasised and published from a central joint authoritative source. Hence implicit doctrine was adopted as a substitute and although this proved to be effective by the end of the war British airborne forces took longer to achieve their potential military effectiveness through that process.

Conclusion

Operation VARSITY undoubtedly represents the pinnacle of the military effectiveness of British airborne warfare. It provides a remarkable contrast to those demonstrations for Churchill and the early raids of 1941 and 1942. It also demonstrates the significant progress that was made from Sicily in 1943 through Normandy in 1944 and even in the six months after Operation MARKET GARDEN. What is self-evident is that the wartime environment was a constant condition that continually influenced the entire developmental process. There has been extensive debate over the contrasting results of military innovation during wartime as opposed to peacetime.[1] Britain's airborne forces travelled from inception to apogee during the Second World War; no component of the process occurred in peacetime. Therefore direct influences on the development process were entirely a product of the wartime environment. It can be contended that innovation during war may be a more efficient and effective process. The operational imperative is more obvious and the performance of a new capability and the time to introduce it into service become more important than the cost, ensuring resources are more readily made available. However, as the development of airborne forces demonstrates, no matter how abundant resources may become there will always be competing interests for their apportionment.

This competition for resources means there has to be fast and visible evidence that the investment is producing a militarily effective return. This fact had considerable influence on the early development of Britain's airborne forces as units had to be committed to raiding operations in order to justify their existence. This caused limited attrition of resources, considerable diversion of staff effort away from the development process, and the adoption of a concept and doctrine that later impinged on mainstream formation level operations. Notwithstanding this the fact that the fledgling airborne establishment competed at all against the momentum of the bomber offensive prior to mid-1943 is a creditable achievement. This at least allowed British airborne forces to survive, if only

CONCLUSION

just, to a point where a clear operational imperative forced resistance to reduce. Limitations in resources continued have an effect until the end of the war and clearly the disaster at Arnhem was at least in part directly due to the lack of available aircraft. However, by Operation VARSITY the airborne establishment had learned to cope with these limitations through thorough and pragmatic planning at a joint and multi-national level.

The second effect of wartime innovation in the case of British airborne forces was the influence that the environment and events had on doctrinal development. As has been shown, the dearth of a centrally endorsed concept and doctrine caused the vacuum to be filled by the thoughts and experience of commanders and practitioners. This was not necessarily a detrimental or retrograde course, rather an expedient means of producing the guidance required by those involved in airborne warfare. However, the drawback was that successive iterations of the semi-formalised doctrine that resulted were unduly influenced by the unique circumstances of particular operations. The pace of successive operations meant that doctrine, tactics and procedures were subject to reiterative review and amendment more often than would have been the case in peacetime. The operational experience gathered by practitioners produced outcomes that differed from the original vision of the developers and resulted in a contorted evolutionary path that sometimes even retarded successful development. At some points in the process this directly translated into a reduction in potential military effectiveness.

British airborne forces were not unique in this trait. The wider publication of doctrine during the Second World War suffered from two inherent problems: timeliness and volume. It has been suggested that when it came to publishing doctrine, a common fault of the British Army throughout the war was that it never managed to collect together what was good while fully excluding what was poor.[2] The problem with much of the published airborne doctrine was that it collected what had been good but was not necessarily relevant for the future. This can be seen most clearly in ATI No.5 with its reproduction of German doctrine that had been successful a year earlier, a year during which the military environment in Europe had changed significantly. However it is to the credit of the General Staff that it managed to publish any airborne doctrine at all in 1941. ATI No.5 was only eight pages long but its publication was still a minor achievement considering the lack of clear conceptual thought and practical experience available at that time. There was a deliberate effort to update published airborne doctrine throughout the war with 'Airborne Operations Pamphlet No.1' in 1943 and then 'Army/Air Operations Pamphlet No.4 – Airborne/Air Transported Operations' produced in early 1945.[3] The fact that the material contained within these official publications was largely obsolete by the time it reached the hands of the average staff officer is possibly irrelevant. Whether any of the wider Army published doctrine was effective or even heeded in some cases is questionable. The plethora of doctrine and instructions on all subjects published by the War Office often defeated their own purpose through their very volume. They were often left unread or only skimmed, and the excess of pamphlets confused many staff officers, not knowing where to look for unfamiliar information. In all probability published airborne doctrine suffered in a similar manner.[4]

There were several factors apparent during airborne development that it would appear could have been improved by the intervention of strong, centrally focused leadership: for example the procurement of aircraft and the provision of glider pilots. Much has been written concerning the role of the individual in military innovation and many studies conclude that some type of unique personality is required to provide leadership and expedite the process of development. It has been reasoned that military 'mavericks' are required to translate the innovation directed by civilian leaders into changes in military doctrine.[5] In order to visualise that change it has been suggested that a degree of 'unconventional creativity'[6] is required, what General Sir John Burnett-Stuart, the interwar proponent of mechanised warfare called 'a touch of divine fire'.[7] However, these theories are over simplistic and do not take account the many different requirements of the development process.

Here airborne development has been studied through the examination of separate lines of development. In doing so it is possible to discern three broad levels to the development process. First is the impetus to innovate: the identification of the requirement for a new military capability and the means to express that requirement to the wider military establishment in terms that were easily understood. Second is the impulse to change: the recognition, implementation and scrutiny of the major changes needed to accommodate the new capability. Third is the drive to develop: the minutiae of the process, the day-to-day enthusiasm and dedication needed to carry innovation forward. These levels of development were not discrete and distinct but overlapped and had greater or less prominence at different points during the process. Creating an impetus to innovate, providing the impulse to change, or driving grass roots development all require different approaches and styles of leadership. Therefore a range of individuals with a variety of personal characteristics and skills are required during different stages of the development process.

It has been demonstrated that although the Air Ministry and War Office were investigating airborne forces prior to 20 June 1940, it was Churchill who provided the impetus to initiate actual development. He had the vision to recognise the potential and future requirement for airborne forces and express that view, albeit imperfectly, to the military establishment. However, implicit in providing the impetus to innovate is ensuring the maintenance of that impetus and here Churchill's contribution was weak

Conclusion 203

and inconsistent. This is easily excusable given his list of priorities, particularly up until the end of 1942. However he then exacerbated the consequences of his intermittent leadership by failing to delegate and identify and appoint anyone capable of developing the impulse to change. This was the predominant factor in shaping the developmental path of British airborne forces during the Second World War and ultimately their military effectiveness during operations. A truly joint permanent headquarters or committee established at least at three star level, with a strong and empowered leader in control was the single act that could have led to fundamental improvement in progress across the lines of development. It has been suggested that Amery could have fulfilled this vital leadership role.[8] Amery, however, despite his pre-war association with projecting military force by air and his obvious interest in airborne forces in India, had developed an antagonistic relationship with the Air Ministry that would ultimately have been counterproductive. Admittedly it is not easy to identify a suitable candidate in hindsight and perhaps this is in part why no one was appointed. Certainly there was no one from within the airborne establishment that fitted the criteria. Browning came closest and his enthusiasm and political awareness would have been assets but he did not reach suitable rank until April 1944, and akin to Amery, his personality hindered inter-service and multinational relationships. It may have been judicious to look for a candidate from outside of the Army and RAF, a suitably empowered civilian or perhaps even a naval officer; a Mountbatten-type character perhaps?

Had a suitable leader been identified and appointed his organisation would have had a clear mandate and definite areas that required close and immediate attention. Chief among these was the direction of the procurement of support and transport aircraft in order to ensure a programme of supply sufficient to meet the airborne establishment's requirements for training and operations. Second was the need to create a coherent airborne concept, produce valid joint doctrine and then monitor tactical development and analyse the experience gained during operations. These two subjects were the leading factors in influencing the developmental path and military effectiveness of British airborne forces during the first and second halves of the process.

Without this critical layer of leadership in place much of its potential remit had to be fulfilled by those operating in the final layer, driving development. This tactical level of leadership was in effect collectively responsible for dictating the direction and pace of airborne development. Rock is perhaps one of the few officers linked with airborne development who could have been considered to possess a 'touch of divine fire'. Otherwise, most of the main protagonists were middle to senior ranking, competent staff officers and battlefield commanders. Neither Browning, nor Gale nor Down could be said to have possessed any great degree of 'unconventional creativity' and they certainly could not be classified as mavericks. These were officers who dedicated themselves to the new form of warfare and took time to study the problems associated with it. They were highly effective trainers and administrators and staunch advocates of their chosen area while remaining cognisant of its limitations. These traits were not confined to the Army, as RAF officers such as Wing Commander Maurice Newnham and Squadron Leader Louis Strange did a huge amount of good work in developing both parachute and glider training at Ringway. Certainly in the case of Browning, Gale and Down and others such as Hill and Frost, these were officers able to inspire and drive those numerous and anonymous individuals who were vital to the development process and to create intense personal and collective loyalty. As such they conform to a category of individuals who can promote a new capability and motivate those involved in its development.[9] These were the men who were ultimately responsible for procuring the aircraft, training the soldiers, writing the doctrine and commanding airborne units and formations around the Mediterranean and across northwest Europe.

British airborne development was driven forward by enthusiastic advocacy, a degree of individual innovation and good old-fashioned, meticulous staff work at the tactical level. This 'bottom-up' approach to development, although it has been argued can be the fastest and most effective means of adaptation, was by its very nature fractured and often unfocused.[10] No matter how committed the personnel involved at the tactical level of development, without the critical layer of leadership required to compel the necessary institutional and systemic change and provide coherence, this approach resulted in a developmental path that fluctuated and lurched forward. However, when viewed from inception to apogee the process as a whole has to be considered a success, a fact displayed by the remarkable level of military effectiveness achieved during Operation VARSITY in March 1945.

Bibliography

PRIMARY SOURCES (UNPUBLISHED)

National Archives, Kew, Surrey
Air Ministry Papers
AIR2, Registered Files.
AIR5, Air Historical Branch Papers.
AIR8, Department of the Chief of the Air Staff, Registered Files.
AIR10, Ministry of Defence and Predecessors: Air Publications and Reports.
AIR16, Fighter Command, Registered Files.
AIR19, Private Office Papers.
AIR20, Papers Accumulated by the Air Historical Branch.
AIR23, RAF Overseas Commands: Reports and Correspondence.
AIR29, Operations Record Books, Miscellaneous Units.
AIR32, Flying Training Command and Technical Training Command, Registered Files and Reports.
AIR37, Allied Expeditionary Air Force, later Supreme Headquarters Allied Expeditionary Force (Air), and 2nd Tactical Air Force: Registered Files and Reports.
AIR39, Army Cooperation Command, Registered Files.
AIR47, 333 Group, later Eastern Air Command: Allied Invasion of North Africa (Operation Torch), Planning Papers.
AIR51, Mediterranean Allied Air Forces: Microfilmed Files.
AIR75, Marshal of the RAF Sir John Slessor: Papers.

Ministry of Aviation Papers
AVIA9, Papers from the Private Office of the Ministers and Parliamentary Secretaries of the Ministry of Aircraft Production.
AVIA15, Ministry of Aircraft Production and Predecessor and Successors: Registered Files

AVIA21, Ministry of Aircraft Production and Ministry of Supply: Airborne Forces Establishment, Later Airborne Forces Experimental Establishment: Reports.
AVIA38, Ministry of Supply and Ministry of Aircraft Production: North American Supply Missions, Second World War.

Board of Trade Papers
BT28, Ministry of Production: Correspondence and Papers.

Cabinet Office Papers
CAB21, Registered Files.
CAB65, Minutes.
CAB66, Memoranda.
CAB79, Chiefs of Staff Committee, Minutes.
CAB 80, Chiefs of Staff Committee, Memoranda.
CAB84, Joint Planning Committee, later Joint Planning Staff, and Sub-committees: Minutes and Memoranda.
CAB98, War Cabinet and Cabinet: Miscellaneous Committees: Minutes and Papers.
CAB106, Archivist and Librarian Files.
CAB120, Ministry of Defence: Secretariat.
CAB121, Special Secret Information Centre.
CAB123, Office of the Lord President of the Council: Registered Files, Correspondence and Papers.

Ministry of Defence Papers
DEFE2, Combined Operations Headquarters, and Ministry of Defence, Combined Operations Headquarters later Amphibious Warfare Headquarters: Records.

Special Operations Executive Papers
HS6, Western Europe.

Prime Minister's Office Papers
PREM3, Operational Correspondence and Papers.
PREM4, Confidential Correspondence and Papers.

Ministry of Supply Papers
SUPP 14, Ministry of Supply General Files.

Treasury Papers
T161, Supply Department: Registered Files.
T162, Establishments Department: Registered Files.
T196, Exchange Requirements Committee: Minutes and Papers.
T246, Central Priority Department: Correspondence, Papers and Minutes.

BIBLIOGRAPHY

War Office Papers
WO32, Registered Files (General Series).
WO106, Directorate of Military Operations and Military Intelligence, and Predecessors: Correspondence and Papers.
WO 107, Quarter Master General's Department: Correspondence and Papers.
WO 171, Allied Expeditionary Force, North West Europe (British Element): War Diaries.
WO 172, British and Allied Land Forces, South East Asia: War Diaries.
WO185, Ministry of Supply: Registered Files.
WO193, Directorate of Military Operations and Plans, later Directorate of Military Operations: Files concerning Military Planning, Intelligence and Statistics (Collation Files).
WO199, Home Forces: Military Headquarters Papers, Second World War.
WO201, Middle East Forces; Military Headquarters Papers, Second World War.
WO203, South East Asia Command: Military Headquarters Papers, Second World War.
WO204, Allied Forces, Mediterranean Theatre: Military Headquarters Papers.
WO205, 21 Army Group: Military Headquarters Papers, Second World War.
WO216, Office of the Chief of the Imperial General Staff: Papers.
WO218, Special Services War Diaries, Second World War.
WO219, Supreme Headquarters Allied Expeditionary Force: Military Headquarters Papers, Second World War.
WO233, Directorate of Air: Papers.
WO291, Military Operational Research Unit, successors and related bodies: Reports and Papers.

Defence Academy Library, Watchfield
Original Papers
OVERLORD – Appreciation of Situation by Brigadier S.J.L. Hill DSO MC, undated.

Battlefield Tour Documentation
Staff College Battlefield Tour 1950 – 6th Airborne Division, OVERLORD.
Staff College Battlefield Tour 1947 – Operation VARSITY.

Imperial War Museum, London
Department of Documents
Address by Lieutenant General Sir Frederick Browning.
Papers of Lieutenant J.W.H. Blower.
Papers of Captain W.C. Brown.

Papers of Lieutenant General Sir Napier Crookenden KCB DSO OBE DL.
Papers of P.R. Devlin.
Papers of Reverend E.N. Downing.
Papers of Air Commodore P.R.C. Groves.
Papers of Brigadier T. Haddon CBE.
Papers of Air Chief Marshal Sir Leslie Hollinghurst.
Papers of D.A. Kerven.
Papers of General Sir Gerald Lathbury GCB DSO.
Papers of Captain R. Marshall.
Papers of Major A.A.K. Pope.
Papers of Brigadier P.N.R. Stewart-Richardson MBE.
Papers of Brigadier A.G. Walch OBE.

Department of Printed Books
1st Airborne Divisional Engineers, Operation FRESHMAN.

Liddel Hart Centre for Military Archives, King's College, London
Papers of Field Marshal Lord Alanbrooke.
Papers of Brigadier J. Drummond.
Papers of General Sir J.W. Hackett.
Papers of Brigadier Sir M.C.A. Henniker.
Papers of Captain Sir B.H. Liddell Hart.
Papers of Major General G. Lloyd.
Papers of D. Russell.

Airborne Forces Museum, Aldershot
1/15, Policy for the Employment of Airborne Forces, Chiefs of Staff Papers, September 1943–March 1944.
1A/2, Ringway Training, 1941–1945.

British Army Tactical Doctrine Retrieval Cell
——, *Airborne Operations, A German Appraisal* (Washington: United States Army Department, 1951).

Museum of Army Flying
1 Airborne Division Report on Operation Market Garden, 10 January 1945.

PRIMARY SOURCES (PUBLISHED)

Official Publications
——, *Airborne Operations Pamphlet No.1 – General*, 1943.
——, *Army/Air Operations Pamphlet No.4 – Airborne/Air Transported Operations*, 1945.

―――――, *Army Training Instruction No.5, Employment of Parachute Troops*, 1941.

―――――, *Bomber Command: The Air Ministry Account of Bomber Command's Offensive Against the Axis, September 1939–July 1941* (London: HMSO, 1941).

―――――, *By Air to Battle: The Official Account of the British Airborne Divisions* (London: HMSO, 1945).

―――――, *Combined Operations 1940–1942* (London: HMSO, 1943).

―――――, *Military Training Pamphlet No.50, Airborne Troops,* August 1941.

―――――, *Report by the Supreme Commander to the Combined Chiefs of Staff on the Operations in Europe of the Allied Expeditionary Force, 6 June 1944 to 8 May 1945* (London: HMSO, 1946).

―――――, *Army Doctrine Publication, 'Land Operations'* (London: MOD, 2005).

―――――, *The Provenance, Authority and Coherence of Concepts* (London: MOD, 2003).

MEMOIRS, etc.

Alexander, H.R.L.G., *The Memoirs of Field Marshal Earl Alexander of Tunis 1940–1945* (London: Cassell, 1962).

Anderson, D., *Three Cheers for the Next Man to Die* (London: Hale, 1983).

Andrews, H. N., *So You Wanted to Learn to Fly, Eh?* (Burnaby: Simon Fraser University, 1997).

Bankhead, H., *Salute to the Steadfast* (London: Ramsay Press, 2002).

Blockwell, A., *Diary of a Red Devil* (Solihull: Helion, 2005).

Carling, H., *Not Many of us Left* (Hailsham: J&KH Publishing, 1997).

Chatterton, G., *The Wings of Pegasus* (London: MacDonald, 1962).

Churchill, W.S., *The Second World War: Volume II, Their Finest Hour* (London: Cassell, 1949).

Dank, M., *The Glider Gang* (London: Cassell, 1977).

De Guingand, F., *Operation Victory* (London: Hodder & Stoughton, 1947).

Deane-Drummond, A., *Arrows of Fortune* (London: Leo Cooper, 1992).

Deane-Drummond, A., *Return Ticket* (London: Collins, 1953).

Edwards, D., *The Devil's Own Luck* (London: Leo Cooper, 1999).

Eisenhower, D.D., *Crusade in Europe* (London: Heinemann, 1948).

Frost, J., *A Drop Too Many* (London: Leo Cooper, 1980).

Frost, J., *Nearly There: The Memoirs of John Frost of Arnhem Bridge* (London: Leo Cooper, 1991).

Gale, R., *With the 6th Airborne Division in Normandy* (London: Samson Low, 1948).

Gale, R., *Call to Arms* (London: Hutchinson, 1968).

Gavin, J.M., *On to Berlin* (New York: Bantam, 1979).

Golden, L., *Echoes from Arnhem* (London: Kimber, 1984).

Hagen, L., *Arnhem Lift* (London: Leo Cooper, 1993).

Hollis, L., *One Marine's Tale* (London: Andre Deutsch, 1956).
Hollis, L. and Leasor, J., *War at the Top* (London: Michael Joseph, 1959).
Horrocks, B.G., *A Full Life* (London: Collins, 1960).
Horrocks, B.G., *Corps Commander* (London: Sidgwick & Jackson, 1977).
Ismay, H.L., *The Memoirs of Lord Ismay* (London: Heinemann, 1960).
Jefferson, A., *Assault on the Guns Of Merville* (London: John Murray, 1987).
Kennedy, J., *The Business of War* (London: Hutchinson, 1957).
Miller, V., *Nothing is Impossible* (Kent: Spellmount, 1994).
Montgomery, B.L., *The Memoirs of Field Marshal Viscount Montgomery of Alamein* (London: Collins, 1958).
Montgomery, B.L., *Normandy to the Baltic* (London: Hutchinson, 1947).
Newnham, M., *Prelude to Glory* (London: Sampson Low, 1948).
Pine-Coffin, R.G., *The Tale of Two Bridges* (Petworth: Pine-Coffin, 2003).
Poett, N., *Pure Poett, The Autobiography of General Sir Nigel Poett* (London: Leo Cooper, 1991).
Powell, G., *The Devil's Birthday* (London: Leo Cooper, 2001).
Riley, N., *One Jump Ahead* (London: John Clare, 1984).
Slessor, J., *The Central Blue* (London: Cassell, 1956).
Urquhart, R.E., *Arnhem* (London: Cassell, 1958).
Waldren, A., *Pacifist to Glider Pilot* (Bognor Regis: Woodfield, 2001).
Wavell, A., *Generals and Generalship* (London: The Times, 1941).
Wilkinson, P., *The Gunners at Arnhem* (Norhampton: Spurwing, 1999).

SECONDARY SOURCES (UNPUBLISHED)

Theses

Buckingham, W.F., *The Establishment and Initial Development of British Airborne Forces June 1940 – January 1942* (Ph.D. Thesis, University of Glasgow, 2001).
Harrison-Place, T., *The Tactical Training and Preparation of the British Army for War In Europe, 1939–46*, (Ph.D. Thesis, Leeds University, 1997).
Hart, S.A., *Field-Marshal Montgomery, 21st Army Group and North-West Europe 1944–45*, (Ph.D. Thesis, King's College, University of London, 1995).
Henry, H.G., *The Planning, Intelligence, Execution and Aftermath Of The Dieppe Raid, 19 August 1942.* (Ph.D. Thesis, Cambridge University,1996).
Judkins, P.E., *Making Vision into Power* (Ph.D. Thesis, Cranfield University, 2007).
Peaty, J.R., *British Army Manpower Crisis 1944* (Ph.D. Thesis, King's College, University of London, 2000).

SECONDARY SOURCES (PUBLISHED)

Books

———, *Serve to Lead* (Camberley: RMAS, undated).
Ambrose, S.E., *Pegasus Bridge* (London: Allen & Unwin, 1984).

BIBLIOGRAPHY

Atkinson, R., *An Army at Dawn: The War in North Africa 1942–1943*, New York: Henry Holt, 2002).
Barber, N., *The Day the Devils Dropped In* (London: Leo Cooper, 2002).
Barnett, C., *The Audit of War* (London: Pan, 2001).
Baynes, J., *Urquhart of Arnhem*, (London: Brassey's, 1993).
Beevor, A., *Crete: The Battle and the Resistance* (London: Penguin, 1991).
Bellamy, C., *The Evolution of Modern Land Warfare: Theory and Practice* (London: Routledge, 1990).
Bernage, G., *Red Devils in Normandy* (Bayeux: Heimdal, 2002).
Bloor, F.R., *The Second World War 1939–1945. Army. Royal Electrical and Mechanical Engineers. Volume II – Technical* (London: War Office, 1951).
Bond, B., *Chief of Staff – The Diaries of Lieutenant-General Sir Henry Pownall, Vol 2, 1940–1944*, (London: Leo Cooper, 1974).
Breuer, W., *Operation Dragoon* (Shrewsbury: Airlife, 2002).
Buckingham, W.F., *Paras* (Stroud: Tempus, 2005).
van Buggenum, D., *B Company Arrived* (Renkum: Sigmond, 2003).
Carver, M., *The Seven Ages of the British Army* (London: Wiedenfield & Nicolson, 1984).
Chamberlain, P. and Gander, T., *Infantry, Mountain and Airborne Guns* (London: MacDonald and Janes, 1975).
Cooper, A., *Wot! No Engines – RAF Glider Pilots and Operation Varsity*, (Bognor Regis: Woodfield, 2002).
Crang, J.A., *The British Army and the People's War 1939–1945* (Manchester: Manchester University Press, 2000).
Crookenden, N., *Airborne at War* (London: Ian Allen, 1978).
Crookenden, N., *Dropzone Normandy* (London: Ian Allen, 1976).
D'Este, C., *Decision in Normandy* (London: Penguin, 1983).
Danchev, A. and Topman, D. (ed.), *War Diaries 1939–1945: Field Marshal Lord Alanbrooke* (Los Angeles: University of California Press, 2001).
De Guingand, F., *Generals at War* (London: Hodder & Stoughton, 1964).
Devlin, G., *Silent Wings* (London: W.H. Allen, 1985).
Dover, V., *The Sky Generals* (London: Cassell, 1981).
Ellis, J., *The Sharp End, the Fighting Man in World War II* (London: Pimlico, 1993).
Englander, D. and Mason, T., *The British Soldier in World War Two* (Warwick: Warwick Working Papers In Social History, 1989).
Ferris, J.R., *Men, Money and Diplomacy* (New York: Cornell, 1989).
Flint, K., *Airborne Armour* (London: Helion, 2004).
Foxall, R., *The Guinea-Pigs – Britain's First Paratroop Raid*, (London: Hale, 1983).
Fraser, D., *And We Shall Shock Them* (London: Hodder & Stoughton, 1983).
French, D., *Raising Churchill's Army* (Oxford: Oxford University Press, 2000).
Fullick, R., *Shan Hackett: The Pursuit of Exactitude* (London: Leo Cooper, 2003).

Gilbert, M., *Churchill – a Biography* (London: Park Lane Press, 1979).
Gilbert, M., *Continue to Pester, Nag and Bite – Churchill's War Leadership* (London: Pimlico, 2004).
Glantz, D., *A History of Soviet Airborne Forces* (London: Frank Cass, 1994).
Golley, J., *The Big Drop* (London: Janes, 1982).
Grey, C.G., *Bombers* (London: Faber & Faber, 1941).
Hamilton, N., *Monty: The Making of a General 1887–1942* (London: Hamish Hamilton, 1981).
Hamilton, N., *Monty: The Master of the Battlefield 1942–1944* (London: Hamish Hamilton, 1983).
Harclerode, P., *Arnhem: A Tragedy of Errors* (London: Cassell, 1994).
Harclerode, P., *Go To It! The Illustrated History of 6th Airborne Division* (London: Caxton, 1990).
Harclerode, P., *Para! Fifty Years of the Parachute Regiment* (London: Cassell, 1992).
Harrison Place, T., *Military Training in the British Army 1940–1944, Dunkirk to D-Day* (London: Frank Cass, 2000).
Hart, S.A., *Colossal Cracks, Montgomery's 21st Army Group in Northwest Europe, 1944–45* (Mechanicsburg: Stackpole, 2007).
Hart, R.A., *Clash of Arms, How the Allies Won in Normandy* (Norman: University of Oklahoma Press, 2001).
Harvey A.D., *Arnhem* (London: Stirling, 2001).
Hastings, M., *Bomber Command* (London: Michael Joseph, 1979).
Hemmings, J., *Silent Approach* (Sussex: The Book Guild, 1999).
Hibbert, C., *Arnhem* (Gloucestershire: The Windrush Press, 1998).
Hickey, M., *Out of the Sky* (London: Mills & Boon, 1979).
Holmes, R., *Firing Line*, (London: Pimlico, 1985).
Hough, R., *Mountbatten: Hero of Our Time* (London: Weidenfeld & Nicholson, 1980).
House, J.M., *Combined Arms Warfare in the Twentieth Century* (Kansas: Kansas University Press, 2001).
Hutson, J.A., *Out of the Blue, U.S. Army Airborne Operations in World War II* (West Lafayette: Purdue University Press, 1998).
Isaacson, J.A., Layne, C. and Arquilla, J., *Predicting Military Innovation* (Santa Monica: Rand, 1999).
James, J., *A Fierce Quality* (London: Leo Cooper, 1989).
Keegan, J., *The Mask of Command*, (London: Penguin, 1988).
Keegan, J., (ed), *Churchill's Generals* (London: Cassell, 2005).
Kershaw, R., *It Never Snows in September* (London: Ian Allen, 1990).
Lamb, R., *Churchill as War Leader – Right or Wrong* (London: Bloomsbury, 1991).
Lewin, R., *Churchill as Warlord* (London: Batsford, 1973).
Liddell Hart, B.H., *Thoughts on War* (London: Faber & Faber, 1944).
Liddell Hart, B.H., *The Other Side of the Hill* (London: Pan, 1999).

Lloyd, A., *The Gliders* (London: Leo Cooper, 1982).
Louis, R.W., *In the Name of God, Go!* (New York: Norton, 1992).
Luvaas, J., *The Education of an Army: British Military Thought 1815–1940*, (Chicago: University of Chicago Press, 1965).
MacDonald, C., *The Lost Battle: Crete 1941*, (London: Pan, 2002).
Mason, A., Bergeron, R., & James, A., *Operation Thursday: Birth of the Air Commandos*, (Honolulu: UPP, 1994).
Miksche, F.O., *Paratroops: The History, Organisation and Tactical Use of Airborne Formations* (London: Faber & Faber, 1943).
Middlebrook, M., *Arnhem 1944: The Airborne Battle 17–26 September* (London: Penguin, 1995).
Millar, G., *The Bruneval Raid* (London: Bodley Head, 1974).
Miller, R., *Nothing Less Than Victory, the Oral History of D-Day* (London: Pimlico, 1993).
Millett, A.R. and Murray, W., *Military Effectiveness, Volume III, The Second World War* (Boston: Allen & Unwin, 1988).
Mrazek, J.A., *The Glider War* (London: Robert Hale, 1975).
Murray, W. and Hart Sinnreich, R. eds., *The Past as Prologue*, (Cambridge: Cambridge University Press, 2006).
Murray, W. and Millett, A.R., eds., *Military Innovation in the Interwar Period* (New York: Cambridge University Press, 1996).
Nalder, R.F.H., *The History of British Army Signals in the Second World War* (London: Royal Signals Institution, 1953).
Norton, G.G., *The Red Devils* (London: Leo Cooper, 1971).
Otway, T., *The Second World War 1939–1945, Army: Airborne Forces* (London: Imperial War Museum, 1990).
Pack, S.W.C., *Operation Husky* (Vancouver: David & Charles, 1977).
Parker, J., *The Paras* (London: Metro, 2000).
Parker, H.M.D., *Manpower* (London: HMSO, 1957).
Posen, B.R., *The Sources of Military Doctrine* (London: Cornell, 1984).
Postan, M.M., *British War Production* (London: HMSO, 1952).
Praval, K.C., *India's Paratroopers* (London: Leo Cooper, 1975).
Quarrie, B., *Airborne Assault* (Yeovil: Haynes, 1991).
Richards, D., *Portal of Hungerford* (London: Heinemann, 1977).
Rosen, S.P., *Winning the Next War* (New York: Cornell University Press, 1991).
Ryan, C., *A Bridge Too Far* (London: Hamish Hamilton, 1974).
Saunders, H.St G., *The Red Beret* (London: Michael Joseph, 1950).
Scarfe, N., *Assault Division* (London: Collins, 1947).
Shannon, K. and Wright S., *One Night in June* (Shrewsbury: Airlife, 1994).
Simpkin, R., *Deep Battle: The Brainchild of Marshal Tukhachevskii* (London: Brasseys, 1987).
Sítek, A.E. and Blunt, V., *The Flying Soldier: The Air Requirements of Airborne Forces* (London: Alliance Press, 1944).

Sixsmith, E.K.G., *British Generalship in the Twentieth Century* (London: Arms & Armour Press, 1970).
Smith, C., *The History of the Glider Pilot Regiment* (London: Leo Cooper, 1992).
Steer, F., *Arnhem: The Fight to Sustain* (London: Leo Cooper, 2000).
Taylor, A.J.P., *Beaverbrook* (London: Hamish Hamilton, 1972).
Thompson, J., *Ready for Anything* (London: Weidenfeld & Nicholson, 1989).
Thompson, R.W., *Dieppe at Dawn* (London: Hutchinson, 1956).
Tugwell, M., *Airborne to Battle* (London: Kimber, 1971).
Weeks, J., *Airborne Equipment* (London: David & Charles, 1976).
Whiting , C., *Bounce the Rhine* (London: Guild, 1985).
Wiggan, R., *Operation Freshman* (London: Kimber, 1986).
Wilmot, C., *The Struggle for Europe* (London: Collins, 1952).
Winton, H.R. and Mets, D.R, *The Challenge of Change: Military Institutions and New Realities, 1918–1941* (London: University of Nebraska Press, 2000).
Wood, A., *History of the World's Glider Forces* (London: Patrick Stephens, 1990).
Wood, D.H., *A Noble Pair of Brothers* (————:————, 1996).
Woods, R.B., *A Changing of the Guard: Anglo–American Relations, 1941–1946* (Chapel Hill: The University of North Carolina Press, 1990).
Wright, L., *The Wooden Sword* (London: Elek, 1967).

Articles

————, 'The Bartholomew Committee 1940 Final Report', *The British Army Review*, No. 129, Spring 2002.
Buckle, H., 'Roped into Action', *The Eagle*, Vol No.10, No.2, August 2002, p.25.
Gale, R.N., 'The 6th (British) Airborne Division in France', *Army Quarterly*, Vol. XLIX, No. 2, January 1945, pp. 235–243.
Greenacre, J.W., 'Assessing the Reasons for Failure: 1st British Airborne Division Signal Communications during Operation Market Garden', *Defence Studies*, Vol 4, No. 3, Autumn 2004, pp. 283–308.
Hall, D.I., 'From Khaki and Light Blue to Purple', *Journal of RUSI*, 147, No. 5, October 2002, pp. 78–83.
Hill, S.J.L., 'Operation "Torch"', *Army Quarterly*, Vol LI, No. 2, January 1946, pp. 177–186.
Margy, K., 'Tragino 1941 – Britain's First Paratroop Raid', *After the Battle*, No. 81, August 1993, pp. 8—29.
Mason, R.A., 'Innovation and the Military Mind', *Air University Review*, Vol XXXVII, No. 2, January – February 1986, pp.39–44.
Millet, A.R., Murray, W. and Watman, K.H., 'The Effectiveness of Military Organizations.' *International Security*, Summer 1986, pp. 37–71.

BIBLIOGRAPHY

Montgomery, B.L., 'Morale in Battle: Analysis', *British Army Review*, No. 145, Autumn 2008, pp.79–86.

Nesbit, R., and Wallwork, J., 'No Higher Test of Piloting Skill', *Aeroplane Monthly*, Special Supplement, May 1994.

Slessor, J., 'Some Reflections on Airborne Forces', *Army Quarterly*, Vol. LVI, No. 2, July 1948, pp. 161–166.

Stafford, D., 'The Detonator Concept: British Strategy, SOE and European Resistance after the Fall of France', *Journal of Contemporary History*, Vol. 10, No. 2, April 1975, pp. 185–217.

Thompson, P.W., 'How the Germans Took Fort Eben Emael', *United States Army Infantry Journal*, August 1942, pp. 22–28.

Wallwork, J.H., '...No Higher Test of Piloting Skill.', *Supplement to Aeroplane Monthly*, May 1994, pp. 14–22.

Winstanley, D., 'How Critical was Air Power in the Failure of Operation Market Garden?', *Royal Air Force, Air Power Review*, Vol. 7, No. 3, Autumn 2004.

Notes

Chapter One
1. T.B.H. Otway, *The Second World War 1939–1945, Army: Airborne Forces* (London: Imperial War Museum, 1990), pp. 304–306.
2. H.L. Ismay, *The Memoirs Of Lord Ismay* (London: Heinemann, 1960), p. 224.
3. D.M. Glantz, *A History of Soviet Airborne Forces*, (London: Frank Cass, 1994), p. 86.
4. J. Frost, *Nearly There: The Memoirs of John Frost of Arnhem Bridge*, (London: Leo Cooper, 1991), p. 70.
5. NA, WO 32/4371, Daily Telegraph, 26 October 1935.
6. NA, WO 32/4371, Minute DMO&I to DMT, 26 October 1935.
7. For a full examination of the progress in airborne development made by other nations prior to the Second World War see Otway, *Airborne Forces*, pp.6–20 and W.F. Buckingham, *The Establishment And Initial Development Of British Airborne Forces June 1940 – January 1942* (Ph.D. thesis, University of Glasgow, 2001), pp. 19–30.
8. NA, WO 32/4371, Air Ministry to War Office, 25 November 1936.
9. C. Bellamy, *The Evolution of Modern Land Warfare: Theory and Practice* (London: Routledge, 1990), p. 89.
10. NA, WO 32/4371, Deputy CIGS to DMO&I, DMT and DSD, 28 February 1938.
11. J.R. Peaty, *British Army Manpower Crisis 1944* (Ph.D. thesis, King's College, University of London, 2000), pp. 104–112.
12. J. Slessor, *The Central Blue* (London: Cassell, 1956), p. 665.
13. M. Carver, *The Seven Ages of the British Army* (London: Wiedenfield & Nicolson, 1984), p. 287.
14. Figures extracted from D.H. Wood, *A Noble Pair of Brothers* (____:____, 1996), p. 43 and C. Smith, *The History Of The Glider Pilot Regiment* (London: Leo Cooper, 1992), pp. 55–66.
15. Otway, *Airborne Forces*, pp. 123 & 130.
16. NA, WO 204/1818, Report on Airborne Operations – "HUSKY", 24 July 1943.
17. Otway, *Airborne Forces*, p. 287.
18. A.R. Millet, W. Murray and K.H. Watman, 'The Effectiveness of Military Organizations.' *International Security*, Summer 1986, pp. 37–71.
19. J. Slessor, 'Some Reflections on Airborne Forces', *Army Quarterly*, Vol. LVI, No. 2, July 1948, p. 161.

NOTES

Chapter Two
1. H.L. Ismay, *The Memoirs Of Lord Ismay* (London: Heinemann, 1960), p. 116.
2. J. Kennedy, *The Business Of War* (London: Hutchinson, 1957), p. 61.
3. J. Slessor, 'Some Reflections on Airborne Forces', *Army Quarterly*, Vol. LVI, No. 2, July 1948, p. 162.
4. A. Danchev, 'Dill', in J. Keegan (ed), *Churchill's Generals* (London: Cassell, 2005), p. 55.
5. NA, CAB 120/262, Churchill to War Office, 22 June 1940. William Buckingham has correctly identified minor staff activity linked with the possible formation of a parachute force between Churchill the War Office and the Air Ministry earlier in June 1940. However the executive order that initiated development remains the minute of 22 June 1940. Buckingham, *The Establishment And Initial Development Of British Airborne Forces*, pp. 88–95.
6. NA, AIR 32/2, The Provision of an Airborne Force, 25 June 1940.
7. D. Stafford, 'The Detonator Concept: British Strategy, SOE and European Resistance after the Fall of France', *Journal of Contemporary History*, Vol. 10, No. 2, April 1975, p. 192.
8. ibid., p. 198.
9. NA, DEFE 2/791, The Employment of Airborne Troops, 4 September 1940.
10. NA, AIR 75/45, Air Ministry to CLE, 12 August 1940.
11. NA, CAB 120/262, DCO to Prime Minister, 27 July 1940.
12. NA, CAB 120/262, 250th Meeting of the COS Committee, 6 August 1940.
13. NA, AIR 2/7470, Summary of Decisions Concerning Airborne Forces, 28th April 1941.
14. NA, CAB 120/262, Prime Minister to Ismay, 1 September 1940.
15. NA, CAB 120/262, Ismay to Prime Minister, 9 September 1940.
16. Otway, *Airborne Forces*, p. 30.
17. NA, AIR 2/7470, Director Military Cooperation (DMC), 2 May 1941.
18. NA, CAB 120/262, Prime Minister's Personal Minute, 27 May 1941.
19. NA, AIR 2/7470, CAS, Army Air Requirements, July 1941.
20. NA, AIR 2/7470, COS Committee Joint Memoranda, Airborne Forces, 29 May 1941.
21. NA, AIR 2/7470, ibid.
22. Otway, *Airborne Forces*, p. 49.
23. NA, AVIA 9/29, Secretary of State Air Ministry of Aircraft Production, 21 March 1942.
24. Otway, *Airborne Forces,*, p. 51.
25. G. Chatterton, *The Wings Of Pegasus* (London: MacDonald, 1962), p. 32.
26. NA, CAB 120/262, The Present Situation of the Airborne Division, 16 April 1942.
27. Otway, *Airborne Forces*, p. 52. Otway is incorrect with this date, as MAP had formed the committee on 4 April 1942.
28. NA, AVIA 9/29, Llewellin to Sinclair, 4 April 1942.
29. NA, AVIA 9/29, Llewellin to Sinclair, 10 April 1942.
30. Ismay, *Memoirs*, p. 159.
31. R. Lewin, *Churchill as Warlord* (London: Batsford, 1973), p.243.
32. ibid, p. 245.

33. NA, AIR 23/5411, First Meeting of the Indian Airborne Forces Committee, 22 April 1941.
34. NA, AVIA 9/29, Llewellin to Sinclair, 20 April 1942.
35. NA, CAB 121/97, Ismay to Churchill, 24 October 1942.
36. NA, CAB 121/97, COS (42) 434, Airborne Forces, 21 October 1942.
37. NA, CAB 121/97 , Ismay to Gale, 7 November 1942.
38. NA, CAB 120/262, Ismay to Churchill, 9 November 1942.
39. NA, CAB 120/262, Churchill to the COS Committee, 12 November 1942.
40. Ismay, *Memoirs*, p. 160.
41. NA, CAB 121/97, Portal to Churchill, 14 November 1942.
42. NA, CAB 120/262, Churchill to the COS Committee, 17 November 1942.
43. NA, CAB 120/262, Ismay to the Deputy Prime Minister, 1 February 1943.
44. NA, WO 193/788, JP(41) 25, Future Plans, Basic Requirements, 17 January 1941.
45. Otway, *Airborne Forces*, p. 94.
46. NA, CAB 120/262, Extract of Report by General Alexander, 21 July 1943 quoted in minute from Brooke to Churchill, 28 July 1943.
47. NA CAB 120/262, Churchill to Ismay, 29 April 1941.
48. NA, CAB 120/262, Stephenson to Churchill, 16 May 42.
49. Kennedy, *The Business of War*, p. 60.
50. F.O. Miksche, *Paratroops: The History, Organisation And Tactical Use Of Airborne Formations* (London: Faber & Faber, 1943), p. 85.
51. NA, AIR 20/3732, Churchill, WP(40) 352, The Munitions Situation, 3 Sep 1940.
52. D.I. Hall, 'From Khaki and Light Blue to Purple', *Journal of RUSI*, 147, No. 5, October 2002, p. 79.
53. ibid, p. 81.
54. W.S. Churchill, *The Second World War: Volume II, Their Finest Hour* (London: Cassell, 1949), p. 144.
55. Kennedy, *The Business Of War*, p. xv.
56. R. Gale, *Call To Arms* (London: Hutchinson, 1968), p. 90.
57. M. Hastings, *Bomber Command* (London: Michael Joseph, 1979), p. 56.
58. Frost, *Nearly There*, p. 77.
59. Ismay, *Memoirs*, p. 178.
60. Churchill, *Their Finest Hour*, p. 339.
61. Kennedy, *The Business Of War*, p. xv.
62. 'The situation is hopeless and I see no other solution besides the provision of an army air arm' and 'The Air Ministry is now so divorced from the requirements of the army that I see no solution except an army air arm.' Danchev and Topman, (ed.), *Field Marshal Lord Alanbrooke*, p .258, (Diary entries for 18 and 19 May 1942).
63. Hastings, *Bomber Command*, p. 56.
64. Gale, *Call To Arms*, p. 127.
65. ibid., p. 125–126.
66. Ismay, *Memoirs*, p. 178.
67. NA, WO 193/788, COS 40(1057), Future Plans – Basic Requirements, 20 December 1940.
68. NA, AIR 32/2, draft CLE paper, Training and Organisation of Airlanding Troops, July 1940.

NOTES

69. NA, AIR 75/45, Airborne Forces, draft note by CAS, 25 September 1942.
70. NA, CAB 121/97, COS (42) 434, Airborne Forces, 21 October 1942.
71. NA, AIR 32/2, Meeting held at the Air Ministry, 5 September 1940.
72. NA 32/2, ibid.
73. NA, AIR 20/3732, Air Ministry, Director of Plans to VCAS, 16 November 1940.
74. NA, AIR 2/7470, Director Military Cooperation, 17 March 1941.
75. NA, WO 193/788, MO1, 26 December 1940.
76. NA, AIR 32/2, DSD to CLE, 7 February 1941.
77. NA, AIR 32/2, Meeting held at the Air Ministry, 5 September 1940.
78. Frost, *Nearly There*, p. 70. Probably few contemporaries believed in airborne forces as deeply as Frost. His point is perhaps that the hierarchy was sceptical and therefore resources were not allocated at the rate or in the quantity that Frost felt they deserved.
79. NA, AIR 75/45, ACAS (Plans), 11 October 1942.
80. A. Danchev, 'Dill', in J. Keegan, (ed), *Churchill's Generals* (London: Cassell, 2005), p. 60.
81. Kennedy, *The Business Of War*, p. 31.
82. Danchev and Topman (ed.), *Field Marshal Lord Alanbrooke*, pp. 172 and 338, (Diary entries for 8 January and 3 November 1942).
83. NA, DEFE 2/791, Airborne Forces, DMC, 9 December 1940.
84. NA, AIR 75/45, Slessor to CLE, 12 August 1940.
85. Gale, *Call To Arms*, p. 127.
86. NA, AIR 2/7470, Nye to Goddard, 7 February 1941.
87. Otway, *Airborne Forces*, p. 59.
88. R. Hough, *Mountbatten: Hero Of Our Time* (London: Weidenfeld & Nicholson, 1980), p. 147.
89. A.R. Millett and M. Murray, *Military Effectiveness, Volume III, The Second World War* (Boston: Allen & Unwin, 1988), p. 108.
90. L. Hollis, and J. Leasor, *War At The Top* (London: Michael Joseph, 1959), p. 119.
91. Millett and Murray, *Military Effectiveness*, p. 108.
92. Hollis and Leasor, *War At The Top*, p. 121.
93. NA, AIR 75/45, Slessor to Strange, 12 August 1940.
94. Millett and Murray, *Military Effectiveness*, p. 109.
95. Hough, *Mountbatten*, p. 151.
96. NA, AIR 2/7470, War Office to Air Minstry, 10 January 1941.
97. Hollis and Leasor, *War At The Top*, p. 10.
98. M.M. Postan, *British War Production*, (London: HMSO, 1952), p. 5.
99. Hollis and Leasor, *War At The Top,,* p. 49.
100. NA, CAB 21/652, Cabinet briefing note, 21 November 1938.
101. A.J.P. Taylor, *Beaverbrook* (London: Hamish Hamilton, 1972), p. 416.
102. Postan, *British War Production*, p. 20.
103. NA, CAB 21/652, ministerial memorandum, 1 November 1938.
104. NA, CAB 21/652, Brief to First Lord of the Admiralty, 2 November 1938.
105. NA, CAB 21/1108, The Creation of the Ministry of Supply, 29 January 1940.
106. See for example Hansard, 17 November 1938, Vol 341, Column 1128–1134.
107. Postan, *British War Production*, p. 137.
108. Hollis and Leasor, *War At The Top*, p. 102.

109. B. Bond, *Chief of Staff – The Diaries of Lieutenant-General Sir Henry Pownall, Vol 2, 1940–1944*, (London: Leo Cooper, 1974), pp. 42 and 48, (Diary entries for 20 September and 20 October 1941).
110. NA, PREM 3/38, Sinclair to Churchill, 1 July 1940. Having been second in command when Churchill commanded the 6th Battalion, Royal Scots Fusiliers, Sinclair was perhaps one of the only men who could criticise Beaverbrook in this manner.
111. NA, AIR 8/480, Beaverbrook to Newall, 10 July 1940.
112. NA, AIR 8/480, Beaverbrook to Sinclair, 11 June 1940.
113. NA, AIR 8/480, Air Ministry, 8 April 1940.
114. NA, AIR 8/480, Relationship of DGE with MAP and the Air Ministry, 28 June 1940.
115. NA, Air 8/480, AMSO, 11 June 1940.
116. NA, AIR 8/480, Relationship of DGE with MAP and the Air Ministry, 28 June 1940.
117. NA, AIR 8/480, Air Ministry, Proposed transfer of DGE, July 1940.
118. NA, AIR 8/480, Sinclair to Beaverbrook, 12 June 1940.
119. NA, PREM 3/38, Newall to Churchill, 11 July 1940.
120. Taylor, *Beaverbrook*, p. 398.
121. ibid, p. 402.
122. NA, PREM 3/38, Beaverbrook to Churchill, 16 June 1940. A.J.P. Taylor presumes that this letter was not sent (Taylor, *Beaverbrook*, p.442). It clearly did reach Churchill whose secretary had noted in the margin that Beaverbrook had requested that the letter be destroyed.
123. NA, PREM 3/38, MAP revised production figures, 11 June 1940. All figures quoted in this section are sourced from this document.
124. Lewin, *Churchill as Warlord* (London: Batsford, 1973), p. 51.
125. NA, PREM 3/38, Air Ministry to MAP, 20 June 1940.
126. NA, PREM 3/38, ibid.
127. NA, PREM 3/38, Beaverbrook to Sinclair, 21 June 1940.
128. NA, PREM 3/38, Sinclair to Beaverbrook, 28 June 1940.
129. Taylor, *Beaverbrook*, p. 431.
130. D. Richards, *Portal Of Hungerford* (London: Heinemann, 1977), p. 205.
131. ibid, p. 467.
132. NA, CAB 120/262, Meeting held in the Air Ministry, 5 September 1940.
133. NA, AIR 20/3732, JPS, Future Plans and Basic Requirements, 18 October 1940.
134. NA, CAB 121/97, COS (41) 366, Formation of an Airborne Brigade in India, 11 June 1941.
135. NA, AIR 23/5932, Records of Operation Analysis, 22 April – 6 June 1941.
136. NA, CAB 121/97, Linlithgow to Amery, 31 March 1942.
137. R.W. Louis, *In the Name of God, Go!* (New York: Norton, 1992), p. 82.
138. Bond, *Chief of Staff*, p. 15, (Diary entry for 25 May 1941). Considering Pownall's assessment of Beaverbrook it is possible that he had a prejudicial low opinion of all politicians (particularly those with Jewish ancestry such as Amery), in which case his value as a commentator might be questioned.
139. NA, CAB 121/97, COS Committee meeting (41)211, 13 June 1941.
140. NA, WO 106/3670, Amery to Dill, 28 October 1941.
141. NA, WO 106/3670, Dill to Amery, 3 November 1941.

NOTES 221

142. NA, CAB 120/262, Amery to Churchill, 6 October 1941.
143. NA, BAB 120/262, Amery to Churchill, 19 January 1942.
144. NA, CAB 120/262, ibid.
145. NA, CAB 120/262, Portal to Churchill, 7 February 1942.
146. NA, CAB 120/262, Amery to Linlithgow, 31 March 1942.
147. NA, CAB 120/262, Amery to Churchill, 9 April 1942.
148. NA, CAB 120/262, Linlithgow to Amery, 18 June 1942.
149. Louis, *In the Name of God, Go!*, p. 151.
150. NA, CAB 120/262, Amery to Churchill, 14 July 1942.
151. Lewin, *Churchill as Warlord*, p .243.
152. M. Gilbert, *Continue to Pester, Nag and Bite – Churchill's War Leadership* (London: Pimlico, 2004), pp. 51 & 73.
153. R. Lamb, *Churchill as War Leader – Right or Wrong* (London: Bloomsbury, 1991), p. 348.
154. Millett and Murray, *Military Effectiveness*, p. 54.

Chapter Three
1. Otway, *Airborne Forces*, pp. 29–30 & 50.
2. NA, T 162/755, establishment of AFEE, undated.
3. Otway, *Airborne Forces*, p. 47.
4. IWM, 99/18/1, the papers of Major A.A.K. Pope, photographs H24569 and H26220. 1,660 German paratroops had been captured during their assault on Holland. Detail of the equipment captured was presented at an Air Ministry conference on 10 June 1940. Buckingham, *Paras,* pp. 68–69.
5. Postan, *British War Production*, p. 104.
6. NA, AIR 39/36, CLE minute, 8 August 1941.
7. Otway, *Airborne Forces*, p. 46.
8. NA, WO 291/445, AORG Report, 24 May 1943.
9. Buckingham, *Paras*, pp. 91–96 and J. Weeks, *Airborne Equipment* (London: David & Charles, 1976), pp. 24–35.
10. NA, T 196/89, Treasury, 27 June 1940 and 4 October 1940.
11. NA, T 196/89, Treasury, 28 September 1940.
12. NA, AIR 39/36, HQ No. 70 Group RAF to HQ AC Comd, 5 August 1941.
13. NA, AIR 20/2385, War Office to Air Ministry, 30 May 1942.
14. NA, AIR 20/4378, MAP Production Summary, 9 October 1942.
15. NA, WO 106/4640, C in C India to Air Min, 5 November 1943.
16. NA, AIR 20/4378, Air Ministry, 27 November 1942.
17. NA, AIR 32/2, Notes on the Use of Parachute and Air-Borne Troops, 4 February 1941.
18. A wicker pannier was also used for airborne resupply that could be pushed from the side door of a Dakota. It was developed by Major Packe after he noted the lightweight and robust quality of his laundry basket while home on leave. F. Steer, *Arnhem: The Fight to Sustain* (London: Leo Cooper, 2000), p. 23.
19. Otway, *Airborne Forces*, pp. 407–409 and Weeks, *Airborne Equipment*, pp. 46–49.
20. NA, AIR 32/2, Notes on the Use of Parachute and Air-Borne Troops, 4 February 1941.
21. NA, AIR 39/36, CLE to HQ No. 70 Group RAF, 2 August 1941.

22. NA, AIR 20/2385, War Office to Air Ministry, 30 May 1942.
23. NA, AIR 2/7470, ACAS to CAS, 24 June 1941.
24. NA, AIR 39/36, Gale to WO, SD4, 7 October 1941.
25. NA, AIR 39/36, AC Comd to Gale, 8 October 1941.
26. NA, AIR 39/36, DMC to AOC-in-C AC Comd, 15 October 1941.
27. NA, AIR 39/38, AOC No. 70 Group RAF to HQ AC Comd, 15 October 1941.
28. J. Frost, *A Drop Too Many*, (London: Leo Cooper, 1980), pp. 172–173.
29. Otway, *Airborne Forces*, pp. 98 & 410–411.
30. French, *Raising Churchill's Army*, p. 88.
31. Frost, *A Drop Too Many*, p. 75.
32. P. Chamberlain and T. Gander, *Infantry, Mountain and Airborne Guns* (London: MacDonald and Janes, 1975), p. 44.
33. P. Wilkinson, *The Gunners At Arnhem* (Norhampton: Spurwing, 1999), p. 23.
34. Chamberlain & Gander, *Airborne Guns*, p. 46.
35. NA, AIR 32/2, CLE to WO, SD4, 15 January 1941.
36. Weeks, *Airborne Equipment*, p. 85.
37. Chamberlain & Gander, *Airborne Guns*, pp. 52–53.
38. Wilkinson, *The Gunners At Arnhem*, p. 19.
39. Otway, *Airborne Forces*, p.174.
40. LHCMA LH 15/8/148, *Army Training Instruction No.5, Employment of Parachute Troops*, 1941, p. 3 & NA, WO 231/126, *Military Training Pamphlet No.50, Airborne Troops*, August 1941, p. 2.
41. Frost, *A Drop Too Many*, pp. 78 & 104.
42. NA, AIR 2/7470, Air Min to WO, 21 May 1941.
43. Otway, *Airborne Forces*, pp.120–121.
44. ibid, p. 46.
45. A. Blockwell, *Diary of a Red Devil* (Solihull: Helion, 2005), pp. 67–68.
46. Wilkinson, *The Gunners at Arnhem*, p. 19.
47. F.R. Bloor, *The Second World War 1939–1945. Army. Royal Electrical and Mechanical Engineers. Volume II – Technical* (London: War Office, 1951), pp. 14–15.
48. NA, AIR 2/7470, Air Min to WO, 21 May 1941.
49. NA, AIR 32/2, summary of airborne brigade group, 26 September 1940.
50. NA, AIR 2/7470, Air Min to WO, 21 May 1941.
51. K. Flint, *Airborne Armour* (London: Helion, 2004), pp. 9–34.
52. NA, AIR 20/2829, Gale to Renwick, 3 September 1942.
53. NA, AIR 20/2829, VCAS to CRD Air Min, 15 September 1942.
54. NA, AIR 20/2829, VCAS to CRD Air Min, 28 September 1942.
55. For a description of the radios commonly used by British airborne forces during the Second World War see L. Golden, *Echoes From Arnhem* (London: Kimber, 1984), pp.139–169.
56. NA, WO 219/5137, Report on Operation MARKET GARDEN, Part III, Index I, December 1944.
57. See as examples R.E. Urquhart, *Arnhem*, (London: Cassell, 1958), p. 47 and Frost, *A Drop Too Many*, pp. 85–87.
58. Otway, *Airborne Forces*, p. 388.
59. R.F.H. Nalder, *The History Of British Army Signals In The Second World War* (London: Royal Signals Institution, 1953), p. 290.

NOTES

60. J.W. Greenacre, 'Assessing the Reasons for Failure: 1st British Airborne Division Signal Communications during Operation Market Garden', *Defence Studies*, Vol 4, No. 3, Autumn 2004, pp.283–308.
61. NA, AVIA 15/573, AFEE to MAP, 22 October 1941.
62. Otway, *Airborne Forces*, pp. 405–406.
63. IWM, K.9996/363, papers of 1 Airborne Divisional Engineers & Wiggan, R., *Operation Freshman* (London: Kimber, 1986), p. 111.
64. Otway, *Airborne Forces*, p. 127.
65. N. Crookenden, *Dropzone Normandy*, (London: Ian Allen, 1976), pp. 189 & 204–205.
66. NA, WO 233/10, British Army Staff, Washington to War Office, 30 November 1943 as an example.
67. NA, WO 233/10, British Army Staff, Washington to War Office, 1 April 1944.
68. NA, AIR 20/2926, VCAS, 21 August 1941.
69. NA, AIR 20/2926, Freeman to Sinclair, 14 April 1942.
70. NA, AIR 2/7338, D Plans, 12 August 1940.
71. Wood, *A Noble Pair of Brothers*, p. 14.
72. NA, AIR 2/7338, D Plans, 12 August 1940.
73. NA, AIR 39/38/1, AOC-in-C No.70 Group, 9 April 1941.
74. Buckingham, *Paras*, p. 102–103 and NA, AIR 20/2926, Aircraft Available in RAF transport squadrons, 21 August 1941.
75. NA, AIR 32/2, Parachute and Airborne Forces, 4 February 1941.
76. NA, AIR 32/2, meeting held at the Air Min, 5 September 1940.
77. NA, AIR 2/7882, COS (42) 142nd meeting, 6 May 1942.
78. NA, AIR 2/7882, meeting held in the Air Ministry, 28 May 1942.
79. NA, CAB 121/97, War Cabinet staff conference, 6 May 1942.
80. NA, CAB 120/262, Keyes to Churchill, 27 July 1940.
81. NA, CAB 121/97, War Cabinet staff conference, 6 May 1942.
82. NA, PREM 3/32/4, Portal to Churchill, 5 October 1942.
83. NA, AIR 20/4078, VCAS to Hollis, 25 May 1940.
84. NA, AIR 20/4078, meeting between Air Min and MAP, 15 May 1940.
85. NA, AIR 19/168, Air Min to MAP, 28 July 1940.
86. NA, CAB 120/262, Keyes to Churchill, 27 July 1940.
87. NA, CAB 120/262, DCO to COS Committee secretariat, 23 August 1940.
88. NA, AIR 39/38/1, DMC to AOC-in-C AC Comd, 7 March 1941.
89. Postan, *British War Production*, p. 341.
90. NA, AIR 2/7566, Air Min DOR to DMC, 5 February 1942.
91. NA, AIR 39/83, correspondence from 29 June 1942 to 2 February 1943.
92. NA, AIR 2/7566, ibid.
93. NA, AIR 21/57. This file contains approximately fifty pages of calculations, diagrams and tables pertaining to just one modification for the Whitley.
94. NA, AIR 39/82, CLE, 12 May 1941.
95. NA, AIR 39/82, Air Min DOR to AOC-in-C AC Comd, 18 June 1942.
96. NA, AIR 39/83, Commander 1 Parachute Brigade, 20 July 1942.
97. NA, AIR 39/83, Air Min DOR to AOC-in-C AC Comd, 8 November 1942.
98. NA, AIR 39/82, DMC to AOC-in-C AC Comd, 15 July 1941.
99. Kenneth Frere, letter to the author, 19 May 2005. Mr Frere was an RAF pilot flying with 38 Group RAF and confirms that during this period he was employed on SOE operations, leaflet drops and tactical bombing sorties.

100. Otway, *Airborne Forces*, pp. 29–30.
101. NA, T 162/755, establishment of CLE, August 1941.
102. NA, T 162/755, establishment of AFEE, undated.
103. NA, AVIA 15/1530, ACAS Technical to Air Ministry, 18 February 1942.
104. NA, AVIA 15/1530, Air Ministry DTD to Air Ministry CRD, 18 February 1942.
105. NA, AIR 2/7566, ACAS Technical to AOC-in-C AC Comd, 23 December 1942.
106. NA, AVIA 15/1530, ACAS Technical to CLE, 21 February 1942.
107. NA, AIR 2/7566, meeting held at AFEE, 18 November 1941.
107. NA, AIR 2/7566, Investigation of Requirements, DMC to War Office (Air 2), 28 October 1942.
109. NA, AIR 2/7566, meeting to discuss airborne forces requirements, 5 December 1942.
110. NA, CAB 121/97, COS(42) 122nd meeting, 17 April 1942.
111. NA, AIR 2/7470, ACAS (Policy), 24 April 1942.
112. NA, CAB 120/262, Keyes to Churchill, 27 July 1940.
113. NA, AIR 2/7338, D Plans, 12 August 1940.
114. NA, AIR 2/7470, Air Ministry, DOR, 25 July 1941.
115. NA, AIR 2/7566, Air Ministry Research and Development, 10 October 1941.
116. NA, AVIA 15/1288. File contains extensive notes on the modifications required for both the Liberator and the Hudson as the result of trials conducted between October and the end of December 1941. In these notes it is suggested that the Liberator might have a capacity of sixteen paratroops but there is no evidence that this was achieved.
117. NA, AIR 19/168, British Consul General, New York to MAP, 27 July 1940.
118. Lamb, *Churchill as War Leader*, pp.77–78.
119. NA, AIR 20/2905, British Air Commission (BAC) to MAP, 29 March 1941. Although this figure appears low 147 Dakotas would equate to a lift of 3,675 paratroopers, in excess of the War Office's requirement for July 1942.
120. Slessor, *The Central Blue*, p. 328.
121. NA, CAB 121/97, COS(42) 142nd Meeting, 6 May 1942 and NA, CAB 120/262 COS Committee memorandum by CAS, 3 May 1942.
122. NA, AVIA 15/1530, summary of technical development on airborne forces requirements, 25 June 1942.
123. NA, AIR 39/83, trials carried out on the C-47 Dakota by AFEE.
124. Otway, *Airborne Forces*, pp. 74–81.
125. NA, AIR 8/916, MAP, aircraft deliveries from USA, 3 July 1941.
126. As an example see Smith, *The Glider Pilot Regiment*, p. 29.
127. Figures extracted from Steer, *Arnhem: The Fight To Sustain*, pp. 93–136.
128. Figures extracted from Wood, *A Noble Pair of Brothers*, pp. 68–70.
129. Otway, *Airborne Forces*, p. 266.
130. In the event the Wellington was not successful as a glider tug as its geodetic design led to a tendency for the aircraft to stretch in flight. Professor Robin Higham, letter to the author, 21 January 2009.
131. NA, AIR 2/7470, Air Min DOR, 25 July 1941.
132. NA, CAB 121/97, CIGS, Airborne Forces, 15 August 1942.
133. Wood, *A Noble Pair of Brothers*, pp. 68–70.
134. NA, AIR 2/7470, Air Ministry, Towing 25-Seater Glider, 12 April 1941.
135. NA, PREM 3/32/4, Churchill to CIGS and CAS, 30 September 1942.

136. NA, PREM 3/32/4, CIGS to Churchill, 3 October 1942.
137. NA, PREM 3/32/4, CAS to Churchill, 5 October 1942.
138. NA, PREM 3/32/4, MAP to Churchill, 27 October 1942.
139. Chatterton, *The Wings Of Pegasus*, p.133.
140. NA, AVIA 15/1530, summary of technical development on airborne forces requirements, 25 June 1942.
141. NA, AVIA 15/1530, MAP Airborne Forces Programme Monthly Progress Report, 30 November 1942.
142. Figures extracted from Wood, *A Noble Pair of Brothers*, p. 43 and Smith, *The History Of The Glider Pilot Regiment*, pp. 55–66.
143. NA, PREM 3/32/5, Churchill's note on minute Portal to Churchill, 11 August 1943.
144. NA, CAB 120/262, Extract from report by General Alexander, 21 July 1943.
145. NA, WO 204/4257, Observations from the Sicilian Campaign, 26 August 1943.
146. NA, AIR 2/7566, War Office, Director of Air to Air Ministry, Director Operations, 14 January 1944).
147. Figures extracted from Wood, *A Noble Pair of Brothers*, pp. 60, 63 and 81.
148. Barnett, C., *The Audit of War* (London: Pan, 2001), p. 148.
149. Postan, *British War Production*, pp. 314–315.
150. NA, AIR 20/2926, conclusions of conference held in Air Ministry, 14 January 1942.
151. NA, PREM 3/482/2, Portal to Churchill, 2 January 1942.
152. NA, AIR 2/7882, briefing note to Air Min PUS, 9 October 1944.
153. NA, AIR 5/278, Secretary of State for Air to Air Member for Supply, 24 August 1922.
154. NA, AIR 5/268, Notes on Rhone Glider Trials, September 1922.
155. NA, AIR 32/2, equipment table, 26 September 1940.
156. NA, AIR 23/6110, RAF Intelligence Report, The German Airborne Attack on Crete, 1 November 1941.
157. NA, AIR 2/5411, Air Supply Apparatus, 2 August 1941.
158. NA, AIR 2/7566, Air Ministry, 9 March 1944.
159. NA, WO 106/4640, C-in-C India to War Office, 26 September 1943.
160. NA, AIR 20/3378, conclusions of conference held in the Air Ministry, 10 June 1940.
161. NA, AIR 20/5256, memorandum on the military employment of gliders, 25 July 1940.
162. NA, AIR 20/5256, notes from meeting held in Air Ministry, 5 August 1940.
163. NA, AIR 20/3378, ACAS (Technical) to ACAS (Requirements), 4 September 1940.
164. NA, AIR 2/7470, Goddard to Nye, 7 March 1941.
165. NA, AIR 2/7470, joint Air Ministry/War Office paper, 24 March 1941.
166. NA, AIR 2/7470, The Development of Airborne Forces, 29 April 1941.
167. AIR 2/7551, CLE to HQ AC Comd, 12 September 1941.
168. AIR 32/2, War Office to Air Ministry, 26 October 1940.
169. AIR 2/7338, MAP to ACAS (Technical), 30 January 1941.
170. AVIA 15/1530, summary of technical development of airborne forces requirements, 25 June 1942.
171. V. Miller, *Nothing Is Impossible*, (Kent: Spellmount, 1994), p. x and A. Waldren, *Pacifist To Glider Pilot* (Bognor Regis: Woodfield, 2001), p. 119 as examples.

172. NA, AIR 32/2, War Office to Air Ministry, 26 October 1940.
173. NA, AIR 32/2, CLE comments on paper by D Plans, Air Min, 14 November 1941.
174. NA, AIR 32/2, Rock to War Office, 15 January 1941.
175. NA, AIR 2/7470, MAP to Air Ministry, 21 May 1941.
176. Smith, The History Of The Glider Pilot Regiment, pp. 83–84.
177. NA, AIR 32/2, notes from a conference held at the Air Ministry, 5 September 1940.
178. NA, AIR 2/7470, notes from a conference held at the Air Ministry, 29 July 1941.
179. Smith, *The History Of The Glider Pilot Regiment*, pp. 47–49.
180. ibid, pp. 47–56.
181. Chatterton, *The Wings Of Pegasus*, p. 43.
182. NA, AVIA 15/2645, report by MAP, 13 August 1943.
183. NA, AIR 2/7551, note by MAP, 30 December 1941.
184. NA, CAB 120/262, Defence Department India to WO, 31 March 1942.
185. NA, AIR 2/7470, DMC to D Air, WO, 6 August 1942.
186. NA, AIR 32/2, papers by CLE, 14 November 1940 and 15 January 1941.
187. NA, AIR 2/7566, paper by DMC, 24 October 1941.
188. NA, AIR 32/3, HQ AC Comd to CLE, 4 September 1941.
189. NA, AIR 2/7551, D Plans, Air Min, 13 June 1941.
190. NA, AIR 2/7551, HQ AC Comd to Air Min, 15 September 1941.
191. NA, AIR 2/7551, MAP minute, 24 November 1941.
192. NA, AIR 2/7470, Air Ministry minutes, 19 February 1941.
193. NA, AIR 2/7551, Moore-Brabazon to Sinclair, 16 August 1941.
194. NA, AIR 2/7551, DMC to Air Min, March 1942.
195. NA, CAB 120/262, note by Ismay, 6 May 1942.
196. NA, CAB 120/262, Cripps to Churchill, 22 September 1942.
197. NA, CAB 120/262, Cherwell to Churchill, 7 November 1942.
198. NA, CAB 121/97, Churchill to Ismay, 16 Novemver 1942.
199. NA, WO 233/43, D Air, War Office, 16 November 1943.
200. NA, WO 233/43, D Air, War Office, 12 September 1944.
201. Buckingham, *Paras*, p. 106.
202. NA, CAB 120/262, Air Ministry to Ismay, 29 April 1941.
203. NA, AIR 32/2, CLE to War Office, 30 October 1940 and 27 January 1941.
204. H.N. Andrews, *So You Wanted to Learn to Fly, Eh?* (Burnaby: Simon Fraser University, 1997), p. 41.
205. Chatterton, *Wings of Pegasus*, p. 32.
206. IWM, 99/18/1, diary of Major Pope, HQ 1 Airborne Division.
207. NA, PREM 3/32/4, COS (43) 17, 11 January 1943.
208. Postan, *British War Production*, p. 18.
209. ibid, pp. 129 & 276.
210. ibid, p. 125.
211. NA, WO 233/43, D Air to DCIGS, 12 June 1944.

Chapter Four
1. H.M.D. Parker, *Manpower* (London: HMSO, 1957), p. 216.
2. French, *Raising Churchill's Army*, p. 86.

NOTES

3. Lewin, *Churchill as Warlord*, p. 51.
4. Parker, *Manpower*, p. 106.
5. T. Harrison Place, *Military Training In The British Army 1940–1944, Dunkirk to D-Day* (London: Frank Cass, 2000), p. 172.
6. NA, WO 32/4371, The Formation of Chasseur Parachutist Companies, *La Revue d'Infanterie*, Vol 88, 1 February 1936, pp. 256–267.
7. NA, WO 32/4371, Report on Parachute Troops, Air Attaché Berlin, 30 August 1938.
8. NA, CAB 120/414, Churchill to Ismay, 3 June 1940.
9. NA, WO 32/4723, Formation of Irregular Commandos, 23 June 1940.
10. NA, WO 32/4723, Volunteers for Special Service, 9 June 1940.
11. NA, WO 32/4723, Volunteers for Special Service, 30 June 1940.
12. NA, CAB 120/262, Keyes to Churchill, 27 July 1940.
13. ABFM, File 1A/2, Major A. Cotterell (War Staff Writer), *Parachuting As A Career*, War No. 59, November 1941.
14. Frost, *A Drop Too Many*, p. 88.
15. IWM, Papers of Lieutenant General Sir Napier Crookenden KCB DSO OBE DL, diary entry for 23 March 1945.
16. IWM, Papers of Brigadier P.N.R. Stewart-Richardson MBE, draft pamphlet 'Airborne Operations', undated.
17. N. Barber, *The Day The Devils Dropped In* (London: Leo Cooper, 2002), p. 59 and 64.
18. R. Miller, *Nothing Less Than Victory, The Oral History of D-Day* (London: Pimlico, 1993), p.2 12.
19. NA, WO 231/126, *Military Training Pamphlet No.50, Airborne Troops*, August 1941, p. 2.
20. French, *Raising Churchill's Army*, pp. 55 & 153.
21. M. Newnham, *Prelude To Glory* (London: Sampson Low, 1948), p. 21.
22. NA, AIR 2/7338, Report on the Central Landing School, Ringway, 31 August 1940.
23. NA, AIR 32/2, Training and Organisation of Airlanding Troops, July 1940.
24. French, *Raising Churchill's Army*, p. 56.
25. NA, WO 32/4371, Parachute Group, Employment, Training and Organisation, 19 April 1938.
26. NA, AIR 39/92, Formation of the CLE, November 1940.
27. See for example R.G. Pine-Coffin, *The Tale Of Two Bridges*, (Petworth: Pine-Coffin, 2003), p. 4.
28. IWM, Papers of Lieutenant General Sir Napier Crookenden KCB DSO OBE DL, diary entry for 12 July 1943.
29. N. Poett, *Pure Poett, The Autobiography of General Sir Nigel Poett* (London: Leo Cooper, 1991), pp. 64–65.
30. NA, AIR 23/5411, meeting held on 11 July 1941.
31. Otway, *Airborne Forces*, p. 34.
32. B.G. Horrocks, *A Full Life* (London: Collins, 1960), p. 187.
33. Gale, *Call To Arms*, p. 117.
34. NA, CAB 120/262, Present Position of the Airborne Division, 16 April 1942.
35. NA, WO 205/751, Instructions for all ranks attending parachute courses, 2 December 1943.

36. J.A. Crang, *The British Army and the People's War 1939–1945* (Manchester: Manchester University Press, 2000), pp. 9–10.
37. Gale, *Call To Arms*, p. 115.
38. Frost, *A Drop Too Many*, p. 18 and J. James, *A Fierce Quality* (London: Leo Cooper, 1989), p. 22.
39. An extraordinary example of this phenomena was the fact that one unit, 22 Independent Parachute Company, was made up of 60 per cent of German Jews who had fled Nazi persecution and joined the British Army with assumed names. G. Bernage, *Red Devils In Normandy* (Bayeux: Heimdal, 2002), p. 17.
40. Gale, *Call To Arms*, p. 115.
41. D. Englander & T. Mason, *The British Soldier In World War Two* (Warwick: Warwick Working Papers In Social History, 1989), p. 2.
42. Otway, *Airborne Forces*, p. 47.
43. IWM, Crookenden, diary entry for 16 May 1943.
44. Englander & Mason, *The British Soldier in World War Two*, p. 13.
45. NA, WO 32/9778 meeting to discuss formation of new air battalions, 26 August 1941.
46. IWM, Papers of D.A. Kerven, p. 6.
47. Peaty, *British Army Manpower Crisis 1944*, p. 137.
48. ABFM, 1A/2, Captain Bradish conducted one such lecture tour in England in July 1941.
49. H. Bankhead, *Salute To The Steadfast* (London: Ramsay Press, 2002), p. 27.
50. ABFM, 1A/2, Major A. Cotterell (War Staff Writer), *Parachuting As A Career*, War No. 59 & 60, November and December 1941.
51. Newnham, *Prelude to Glory*, pp. 75–77.
52. ABFM, 1A/2: *Picture Post*, Vol 22, No.12, 18 March 1944, pp. 7–13. *Boy's Own Paper*, Vol 65, No.1, October 1942, pp.4–6. *Flight Magazine*, 30 October 1941, pp. 297–300. Not all propaganda was positive. An short article in Tatler by Wing Commander E.G. Oakley-Beuttler was accompanied by a cartoon of paratroops tangled in their lines, descending upside down and landing on their heads with the caption 'Any more volunteers?' E.G. Oakley-Beuttler, 'A School for Paratroops', *Tatler and Bystander*, January 20 1943.
53. Newnham, *Prelude to Glory*, pp. 310.
54. Otway, *Airborne Forces*, pp. 54, 94 & 108.
55. NA, WO 32/9778, Formation of further parachute battalions, 7 July 1941.
56. NA, WO 32/9778, meeting held to discuss the raising of additional parachute battalions, 23 July 1941.
57. Pine-Coffin, *Tale of Two Bridges*, p. 5.
58. NA, WO 205/751, Formation of 6 Airborne Division, 6 August 1943.
59. A. Jefferson, *Assault On The Guns Of Merville* (London: John Murray, 1987), p. 28.
60. Peter Wilkinson, letter to the author, 23 May 2005.
61. Company Sergeant Major Swanston of 7 Battalion, The King's Own Scottish Borderers recalled, 'A weeding-out process started and for my own company this process turned out to be satisfactory. The method was an easy one. I marched the men in one by one to the medical officer. I was to stand just behind the man to be examined and then I would just nod yes or no to the

NOTES

doctor. This resulting in a positive or negative check.' Blockwell, *Diary of a Red Devil*, p.63.
62. One officer with 1 Airborne Division after Operation MARKET claims that many of the post-Arnhem replacements were recruited direct from Borstals as an alternative to incarceration but there is no primary evidence of this policy. N. Riley, *One Jump Ahead* (London: John Clare, 1984) p. 145.
63. NA, AIR 29/512, CLE Operational Record Book, introduction and entry for 13 July 1940.
64. NA, DEFE 2/791, letter from Slessor to Major General Bourne, DCO, 4 July 1940.
65. Newnham, *Prelude to Glory*, p. 16.
66. Figures for training fatalities are extracted from ABFM 1A/2, Summary of Fatalities at Ringway, undated.
67. Gale, *Call to Arms*, p. 117.
68. Frost, *A Drop Too Many*, p. 34.
69. NA, WO 32/4371, file notes, 13 December 1935.
70. NA, CAB 120/262, D Plans, Air Ministry, 12 August 1940.
71. ABFM 1A/2, note by Capt M. Lindsay, undated.
72. ABFM 1A/2, notes by R.H Levien, undated, Squadron Leader B.J.O. Winfield dated 31 May 1943 and J. Kilkenny, instructor at No.1 PTS, undated. 'Ringing the bell' was the phrase used to describe hitting ones face or head on the aperture on leaving a bomber aircraft.
73. Newnham, *Prelude to Glory*, pp. 21 & 151. Although no disgrace during initial training any refusal to jump from a trained paratrooper attracted a compulsory court martial.
74. NA, CAB 120/262, COS Committee meeting, 6 August 1940.
75. NA, AIR 39/92, HQ AC Comd to Air Ministry, 10 September 1941.
76. NA, AIR 39/83, DMC, Air Ministry to D. Air, War Office.
77. NA, AIR 29/512, CLE Operational Record Book, introduction, undated.
78. ABFM 1A/2, note by A.L. Shepherd, The Suffolk Regiment, instructor at No.1 PTS, undated.
79. NA, AIR 39/92, scheme for the formation of an Army ground training centre for British parachute troops, 9 September 1941.
80. ABFM 1A/2, note by T. Goode, engineer in CLE workshop, undated.
81. Parachutes had been used as a means of escape by battlefield observers posted in tethered balloons during the First World War.
82. Of the 46 fatalities that occurred at No.1 PTS during the war 26 were attributed to somersaulting and subsequent twisting of rigging lines. ABFM 1A/2, summary of fatalities at Ringway, undated.
83. Newnham, *Prelude to Glory*, p. 50.
84. Figures extracted from ABFM 1A/2, summary of fatalities at Ringway, undated.
85. ABFM, 1A/2, Picture Post, Vol 22, No.12, 18 March 1944, p. 10.
86. ABFM, 1A/2, letter from G.D. Thompson, undated.
87. NA, WO 32/9778, formation of further parachute battalions, 7 July 1941.
88. ABFM, 1A/2, notes for information and guidance in connection with No.1 course of instruction on parachuting, 3 November 1941.

89. NA, AIR 2/7470, policy for airborne forces, 17 March 1941 and DEFE 2/791, meeting to discuss parachute training, 19 July 1940.
90. NA, AIR 2/7470, Ismay to Churchill, 29 May 1941.
91. NA, AIR 29/512, CLE Operational Record Book, entry for 13 December 1941.
92. NA, AIR 2/7470, Joint General Staff/Air Staff memorandum on airborne forces, 4 July 1942.
93. NA, AIR 29/512, CLE Operational Record Book, comparing the entries for 13 December 1941, 7 February 1942, 26 August 1943 and 8 April 1944.
94. NA, AIR 29/512, CLE Operational Record Book, entry for 7 October 1944 shows a course strength of just 68 of which 50 passed.
95. NA, WO 204/4195, AFHQ minute, 22 June 1944.
96. NA, AIR 29/512, CLE Operational Record Book, comparing the entries for 13 December 1941, 7 February 1942, 26 August 1943 and 8 April 1944.
97. The total establishment of the Glider Pilot Regiment never rose above 2,500 men throughout the war. Otway, *Airborne Forces*, p. 56.
98. IWM, Papers of Brigadier A.G. Walch OBE, Opening Address by Colonel Chatterton to Glider Pilot Recruits, undated.
99. Otway, *Airborne Forces*, p. 35.
100. IWM, Papers of Brigadier A.G. Walch OBE, Directive on the Training of Glider Pilots by Colonel Chatterton, November 1944, pp. 4–5.
101. The order of the final two qualities appears to be anomalous. NA, AIR 2/7422, notes for examiners of glider pilot applicants, 17 November 1940.
102. NA, AIR 2/7422, training of glider pilots, 15 December 1940.
103. Otway, *Airborne Forces*, p. 36.
104. Chatterton, *The Wings Of Pegasus*, p. 20.
105. NA, WO 32/9845, formation of an Army glider pilot regiment, 31 October 1941.
106. NA, AIR 2/738, glider pilots, 14 December 1940.
107. NA, AIR 2/7422, Air Ministry, 25 November 1941.
108. See for example Andrews, *So You Wanted to Learn to Fly, Eh?*, pp. 15–18 and M. Dank, *The Glider Gang* (London: Cassell, 1977), p. 42.
109. IWM, Papers of Papers of Brigadier A.G. Walch OBE, op cit, p. 6.
110. NA, AIR 32/2, meeting held in the Air Ministry, 5 September 1940.
111. ibid, Air Ministry to CLE, 11 November 1940.
112. Otway, *Airborne Forces*, p. 35.
113. J.H. Wallwork, '...No Higher Test of Piloting Skill.', *Supplement to Aeroplane Monthly*, May 1994, p. 15.
114. NA, AIR 32/2, CLE paper, 14 November 1940. Coxswain was a term briefly adopted by the War Office to describe glider pilots. Although there are recorded examples, the suggestion that the glider pilot might lead his passengers is perhaps taking the argument to its extreme. 'Guide' might be a more accurate term.
115. NA, AIR 2/7338, Air Ministry, 24 November 1940.
116. Dank, *The Glider Gang*, pp. 206–207.
117. Chatterton, *The Wings Of Pegasus*, p. 206.
118. NA, AIR 2/7470, Note on the Development of Airborne Forces, 29 April 1941.
118. NA, AIR 39/38, Air Ministry meeting, 22 August 1941.
120. NA, AIR 2/7470, joint memo on glider-borne forces, 20 May 1940.

NOTES

121. NA, AIR 29/512, CLE Operational Record Book, entry for 23 August 1940.
122. Reg Leach, letter to the author, 4 December 2006. Reg Leach interviewed Squadron Leader Fender for the Manchester Airport Archive. The other two pilot's names are recorded as Peter Davies and Douglas Davie. Their ranks are not recorded.
123. NA, AIR 39/92, note on the formation of the CLE, November 1940.
124. NA, AIR 2/7422, HQ AC Comd to Air Ministry, 11 February 1941.
125. ibid, War Office to Air Ministry, 24 September 1941.
126. A. Cooper, *Wot! No Engines* (Woodfield: West Sussex, 2002) p. 56.
127. Smith, *The History of the Glider Pilot Regiment*, p. 16.
128. Otway, *Airborne Forces*, p. 35.
129. NA, AIR 39/39, Air Ministry meeting, 5 May 1942.
130. ibid, notes of Air Ministry meeting, 25 February 1942.
131. NA, AIR 20/4661, Air Ministry, 3 July 1944 and NA, WO 32/9845, War Office, the formation of Army glider unit, 26 October 1941.
132. NA, AIR 2/4894, War Office, Training of Glider Pilots, 2 July 1942 and Air Ministry, Army Personnel for Training as Glider Pilots, 17 July 1942.
133. NA, AIR 2/7422, Air Ministry, Airborne Forces Training Policy, 24 March 1942.
134. NA, AIR 39/39, Air Ministry meeting, 5 May 1942.
135. NA, WO 216/33, Pritt to Churchill, 9 March 1943.
136. ibid, Sinclair to Churchill, 18 March 1943.
137. NA, CAB 120/262, Air Ministry meeting, 7 September 1940.
138. ibid, Air Ministry note, The Development of Airborne Forces, 29 April 1941.
139. NA, AIR 2/7470, Air Ministry, Action Required for Providing Airborne Force, 5 August 1941. The mortality rate amongst glider pilots was exceptionally high. During Operation HUSKY 57 out of 163 were killed (35%). Operation MARKET 229 killed out of 1,094 (21%). Operation VARSITY 100 killed out 904 (11%). These totals are for killed only and do not account for losses through serious injury or prisoners of war. Figures extracted from C. Smith, *The History of the Glider Pilot Regiment*, pp. 54, 56, 105, 125, 153–162.
140. NA, AIR 2/7470, draft COS paper, Policy for Airborne Forces, 17 March 1941.
141. Dank, *The Glider Gang*, pp. 171 & 206.
142. NA, WO 23/2546, Reasons for Having a Second Pilot in a Glider, 22 August 1944.
143. NA, CAB 120/262, Brief for COS (42) 138th Meeting, 6 May 1942.
144. NA, AIR 2/4894, Glider Pilot Training – Policy, 10 July 1942.
145. NA, WO 216/33, Airborne Forces, COS (43) 206th Meeting, 28 April 1943.
146. NA, AIR 20/4661, Air Ministry meeting, 12 July 1944.
147. The reason for this surplus of aircrew is unclear. RAF operations continued at a high rate during this period so a reduction in activity is an unlikely cause. It could therefore be assumed that the Flying Training Command had become over efficient in producing trained aircrew. NA, AIR 20/4661, Air Ministry, 31 July 1944.
148. NA, AIR 20/4661, Air Ministry, Training of RAF Pilots as Glider Pilots, 6 September 1944.
149. NA, AIR 20/4661, Air Ministry, 30 September 1944.
150. NA, AIR 20/4661, RAF Flying Training Command, 16 December 1944.
151. NA, AIR 20/4661, Air Ministry, 28 February 1945.
152. NA, WO 23/2546, HQ SEAC, 6 January 1945.

153. A. Cooper, *Wot! No Engines – RAF Glider Pilots And Operation Varsity* (Bognor Regis: Woodfield, 2002), p. 46, Wood, *The Glider Pilot Regiment*, p. 126 and Chatterton, *Wings of Pegasus*, pp. 208–212.
154. L. David Brook, letter to the author, 11 May 2005.
155. Harry Howard, letter to the author, 19 September 2005.
156. Chatterton, *The Wings Of Pegasus*, pp. 40 & 105.
157. NA, WO 205/751, Airborne Liaison Report No.11, 13 October 1943.
158. Miksche, *Paratroops*, p. 8.
159. NA, CAB 120/262, Provision of an Airborne Force, 1 November 1940.
160. French, *Raising Churchill's Army*, p. 168.
161. Harrison Place, *Military Training In The British Army 1940–1944*, p. 19.
162. NA, AIR 32/2, Training and Organisation of Airlanding Troops, July 1940 and CLE Comments on Draft Paper by DD Plans, 14 November 1940.
163. NA, AIR 2/7338, CLE, 31 August 1941.
164. NA, WO 32/9778, AOC-in-C AC Comd to CLE, 7 March 1941 CIGS, Army Air Requirements, 30 May 1941.
165. NA AIR 32/3, Notes on the Provision of a Glider-Borne Force, 29 December 1941.
166. NA AIR 38/39, HQ AC Comd to CLE, 30 May 1941.
167. NA, AIR 29/512, CLE Operational Record Book, introduction and entry for 10 March 1941.
168. NA, AIR 39/39, Air Arrangements Required for the Formation of an Airborne Division at Home, 10 November 1941.
169. NA, AIR 39/39, Air Ministry Meeting, 31 March 1942.
170. NA, AIR39/39, AC Comd, Air Arrangements Required for the Formation of an Airborne Division at Home, May 1942.
171. NA, AIR 39/39, Browning to Barratt, AOC-in-C AC Comd, 7 May 1942.
172. NA, AIR 2/7470, Joint General Staff / Air Staff Memorandum, Airborne Forces, 4 July 1942 & 14 July 1942.
173. Wood, *A Noble Pair of Brothers*, pp. 24–51.
174. NA, AIR 37/280, Air Tactics- Airborne Warfare, 26 August 1942.
175. NA, WO 205/751, SHAEF, Troop Carrier – Airborne Combined Training, 20 February 1944.
176. Kenneth Frere, letter to the author, 19 May 2005 and Bill Angell, letter to the author, 30 May 2005. Both Kenneth Frere and Bill Angell were aircrew with 38 Group during the latter half of the war.
177. NA, AIR 32/2, CLE Minute, 18 September 1940.
178. Chatterton, *Wings of Pegasus*, p. 126.
179. Peter Wilkinson, letter to the author, 23 May 2005. Wilkinson was a Royal Artillery command post officer who received his initial experience of landing in a glider with his heavy equipment on the landing zone at Arnhem.
180. NA, AIR 29/512, CLE Operational Record Book, entry for 26 October 1940.
181. NA, AIR 29/513, CLE Operational Record Book, entries for 12 December 1940, 2 January 1941 and 4 April 1941.
182. Newnham, *Prelude to Glory*, p. 35.
183. NA, DEFE 2/791, CLE draft pamphlet 'Landing Drill for Sub-Sections', 7 November 1940.
184. French, *Raising Churchill's Army*, p. 204.

NOTES

233

185. NA, AIR 2/7470, Air Staff, Aircraft for Airborne Division, March 1942.
186. NA, AIR 2/7470, meeting with Prime Minister, 6 May 1942.
187. NA, AIR 2/7470, DMC to ACAS Plans, Formation of Two Independent Parachute Brigades, 6 August 1942.
188. Peter Wilkinson, letter to the author, 23 May 2005.
189. NA, WO 32/4371, Parachute Group, Employment, Training & Organisation, 19 April 1938.
190. ABFM, File 1A/2, Report by Lieutenant Colonel J.C. Kennedy, 15 April 1941.
191. IWM, Papers of Lieutenant General Sir Napier Crookenden KCB DSO OBE DL, diary entry for 9 January 1944.
192. NA, AIR 37/998, Commander Glider Pilot's Report on Operation NEPTUNE, 26 July 1944 and IWM, 01/11/12, The papers of Brigadier A.G. Walch OBE.
193. Golden, *Echoes From Arnhem*, p.148. Unfortunately during Operation MARKET GARDEN the signals officers' foresight was negated by the heavier than anticipated German defence, particularly artillery and mortar fire which cut cable and made its maintenance practically impossible.
194. Jefferson, *Assault On The Guns Of Merville*, p. 30.
195. NA, WO 205/751, HQ 6 Airborne Division, Airborne Liaison Report No.11, 5 October 1943.
196. A model of the Norsk Hydro plant in Vemork, Norway enabled the members of Operation FRESHMAN to 'draw up a plan and an escape route' after the sabotage was complete. R. Wiggan, *Operation Freshman* (London: Kimber, 1986) p. 42. Chatterton recalls the model prior to Operation DRAGOON was 'extremely accurate, giving a good bird's eye view of the landing zone'. Chatterton, *Wings of Pegasus*, p. 160.
197. Chatterton, *Wings of Pegasus*, p. 130.
198. Jim Wallwork, letter to the author, 7 July 2005.
199. NA, CAB 120/262, CLE, Provision of an Airborne Force 1 November 1940.
200. G. Millar, *The Bruneval Raid*, (London: Bodley Head, 1974) p. 156.
201. Pine-Coffin, *The Tale Of Two Bridges*, p. 17.
202. Chatterton, *Wings of Pegasus*, p. 131.
203. NA, AIR 37/998, Commander Glider Pilot's Report on Operation NEPTUNE, 26 July 1944.
204. Wallwork, '...No Higher Test of Piloting Skill.', pp. 14–19.
205. IWM, File 01/11/12, The papers of Brigadier A.G. Walch OBE, A Directive on the Air and Military Training of the Glider Pilot, November 1944.
206. Jim Wallwork, letter to the author, 7 July 2005.
207. K. Margy, 'Tragino 1941 – Britain's First Paratroop Raid', *After the Battle*, No. 81, August 1993, p. 11.
208. Tony Deane-Drummond, letter to the author, 3 July 2005.
209. Jefferson, *Assault On The Guns Of Merville*, p. 22.
210. IWM, Papers of Lieutenant General Sir Napier Crookenden KCB DSO OBE DL, diary entries for 12 June 1944 and 23 March 1945.
211. Peaty, *British Army Manpower Crisis 1944*, pp. 106–107.
212. ibid, pp. 111–112.
213. All casualty figures taken from Otway, *Airborne Forces*, pp. 87, 123, 130, 191, 283 and 319.
214. J. Ellis, *The Sharp End, the Fighting Man in World War II* (London: Pimlico, 1993), p. 234.

215. B.L. Montgomery, 'Morale in Battle: Analysis', *British Army Review*, No. 145, Autumn 2008, pp. 83–85.
216. V. Miller, *Nothing is Impossible* (Staplehurst: Spellmount, 1994), p. 274.
217. Harrison Place, *Military Training In The British Army*, p. 174.
218. French, *Raising Churchill's Army*, p. 55.
219. NA, AIR 2/7470, D Plans to CAS, 24 June 1941.
220. Otway, *Airborne Forces*, pp. 34, 38, 51 & 140.

Chapter Five
1. J. Luvaas, *The Education of an Army: British Military Thought 1815–1940*, (Chicago: University of Chicago Press, 1965), p. 339.
2. 'The moral component is concerned with the least predictable aspect of operations – the human element. If the human element is neglected, the penalties to be paid will be great and battle-losing. If time and effort are invested in the human element, all things become possible.' Anon, *Army Doctrine Publication, 'Land Operations'* (MOD: London, 2005), p. 143.
3. For a detailed examination of the officer selection, training and promotion process during the Second World War see Crang, *The British Army and the People's War1939-1945*, pp. 21–57.
4. R. Holmes, *Firing Line*, (London: Pimlico, 1985), p. 342.
5. J. Keegan, *The Mask of Command*, (London: Penguin, 1988), pp.329–338.
6. See for examples MOD, *Land Operations*, p. 137, Wavell, A., *Generals and Generalship* (London: The Times, 1941), p. 8 and Anon, *Serve to Lead* (Camberley: RMAS, undated), p. 31.
7. LHCMA, Gale to Liddell Hart, 8 April 1952.
8. Chatterton, *The Wings Of Pegasus*, p. 177.
9. NA, WO 285/1, General Dempsey's appreciation of the capture of the area east of the River Orne, 14 February 1944.
10. NA, WO 285/1, The employment of airborne forces, 21 March 1944.
11. NA, WO 285/1, The employment of airborne forces, April 1944.
12. NA, WO 285/1, Co-operation required in planning by 1 Airborne Division and HQ 8 Corps, 9 May 1944.
13. NA, WO 285/1, The employment of airborne forces, 21 March 1944.
14. ABFM, File 1/15, Report to the Combined Chiefs of Staff, Policy as to the Organisation and Employment of Airborne Troops, 6 March 1944.
15. The operational level of command can be defined as 'the employment of substantial land, sea and air forces, either singly or in combination, in a discrete and definable theatre of war over a finite period of time.' J. Gooch, "History and Nature of Strategy" in W. Murray and R. Hart Sinnreich, eds., *The Past As Prologue*, (Cambridge: Cambridge University Press, 2006), p. 141.
16. Otway, *Airborne Forces*, p. 227.
17. ABFM 1/15, Report to the Combined Chiefs of Staff, Policy as to the Organisation and Employment of Airborne Troops, 6 March 1944. In the final minutes the second sentence was deleted.
18. NA, CAB 120/262, Joint Staff Mission to COS Committee, 9 August 1944.
19. Alexander hardly even mentions the airborne forces under his command in his autobiography. H.R.L.G. Alexander, *The Memoirs of Field Marshal Earl Alexander of Tunis 1940–1945* (London: Cassell, 1962).

NOTES

20. NA, CAB 120/262, Report by General Alexander, 21st July 1943 quoted in minute from Brooke to Churchill, 28 July 1943 and PREM 3/32/4, COS (43) 17, 11 January 1943.
21. NA, WO 106/3913, Alexander to Brooke, 21 September 1943.
22. Otway, *Airborne Forces*, p. 449.
23. N. Hamilton, *Monty: The Master Of The Battlefield 1942–1944* (London: Hamish Hamilton, 1983), p. 389.
24. Otway, *Airborne Forces*, pp. 216–222.
25. R. Atkinson, *An Army at Dawn: The War in North Africa 1942–1943*, (New York: Henry Holt, 2002), p. 376.
26. Hamilton, *Monty: The Master Of The Battlefield*, pp. 457 & 460.
27. NA, WO 106/2764, Outline Plans for Operation TORCH, 9 & 21 August 1942.
28. NA, WO 204/4585, Correspondence between Eisenhower and Clark concerning planning for Operation TORCH, 9 & 21 August, 8, 15, 19 & 23 September, 9 October 1942.
29. NA, WO 106/2786, HQ Allied Forces, C in C's Dispatch from the North African Campaign 1942–43.
30. F. de Guingand, *Operation Victory* (London: Hodder & Stoughton, 1947), p. 173.
31. N. Hamilton, *Monty: The Making Of A General 1887–1942* (London: Hamish Hamilton, 1981), pp. 446–459.
32. ibid, pp. 470–471.
33. D.D. Eisenhower, *Crusade In Europe* (London: Heinemann, 1948), p. 188.
34. NA, WO 204/1818, Report on Airborne Operations – "HUSKY", 24 July 1943.
35. Hamilton, *The Master Of The Battlefield*, p. 304.
36. ibid, p. 354.
37. NA, WO 32/16303, Allied Forces HQ, C-in-C's Dispatch, The Sicilian Campaign 1943.
38. NA, WO 204/1818, Allied Forces HQ, Training Memorandum 43, Employment of Airborne Forces, 2 August 1943, p. 1.
39. Hamilton, *Monty: Master Of The Battlefield*, p. 521.
40. ABFM 1/15, Report to the Combined Chiefs of Staff, Policy as to the Organisation and Employment of Airborne Troops, 6 March 1944, p. 1.
41. NA, CAB 120/262, Joint Staff Mission Report 182, 9 August 1944.
42. Leigh-Mallory had already objected to part of the airborne plan for D-Day but on that occasion was over ruled and subsequently proved wrong. Hamilton, *Monty: Master Of The Battlefield*, pp. 634 & 648.
43. Otway, *Airborne Forces*, pp. 206–213. Horrocks even considered apologising to the airborne HQ for the speed of his advance. B.G. Horrocks, *Corps Commander* (London: Sidgwick & Jackson, 1977), p. 76.
44. E.K.G., Sixsmith, *British Generalship In The Twentieth Century* (London: Arms & Armour Press, 1970), p. 260.
45. Eisenhower, *Crusade In Europe*, pp. 423 & 424.
46. Otway, *Airborne Forces*, p. 446.
47. Sixsmith, *Generalship*, p. 193.
48. NA, AIR 2/7470, Joint General Staff/Air Staff Memorandum, 4 July 1942.
49. IWM, Papers of Lieutenant General Sir Napier Crookenden KCB DSO OBE DL, Vol II. A. Deane- Drummond, *Arrows Of Fortune* (London: Leo Cooper, 1992), p. 3, Frost, *A Drop Too Many*, pp. 1–13, James, *A Fierce Quality*, pp. 18–21.

50. French, *Raising Churchill's Army*, p. 20.
51. S.J.L. Hill, 'Operation "Torch"', *Army Quarterly*, Vol LI, No. 2, January 1946, p. 179.
52. IWM, Papers of Lieutenant General Sir Napier Crookenden KCB DSO OBE DL, Vol III, entry for 6 June 1943.
53. French, *Raising Churchill's Army*, p. 161.
54. Otway, *Airborne Forces*, pp. 429–434.
55. Defence Academy Library, OVERLORD – Appreciation of Situation by Brigadier S.J.L. Hill DSO MC, undated.
56. Gale, *Call To Arms*, p. 116.
57. IWM, Papers of Lieutenant General Sir Napier Crookenden KCB DSO OBE DL, Vol III, entry for 19 August 1943.
58. NA, WO 32/9778, meeting held to discuss formation of two new air battalions, 30 August 1941.
59. For a full account of Gale's career see Gale, *Call To Arms*.
60. Poett's autobiography gives a full account of his career, Poett, *Pure Poett*. Hackett's biography gives a full account of his career, R. Fullick, *Shan Hackett: The Pursuit Of Exactitude* (London: Leo Cooper, 2003). A synopsis of Bols's career can be found in V. Dover, *The Sky Generals* (London: Cassell, 1981), pp. 155–172.
61. Poett, *Pure Poett*, p. 53.
62. Dover, *The Sky Generals*, p. 155.
63. Chatterton, *The Wings Of Pegasus*, pp. 38 & 115.
64. IWM, Papers of Lieutenant General Sir Napier Crookenden KCB DSO OBE DL, Vol III, entries for 6 June 1943 and 15 June 1944.
65. Dover, *The Sky Generals*, p. 71.
66. Tony Deane-Drummond letter to the author, 6 July 2006 and Chatterton, *The Wings Of Pegasus*, p. 38.
67. Dover, *The Sky Generals*, p. 73.
68. Chatterton, *The Wings Of Pegasus*, p. 40.
69. Otway, *Airborne Forces*, p. 133.
70. Lathbury had been a candidate to succeed Down as GOC 1 Airborne Division in January 1944 and may even have unofficially and erroneously been informed as such. Urquhart, Arnhem, p. 27 and J. Baynes, *Urquhart*, (London: Brassey's, 1993), p. 72.
71. Dover, *The Sky Generals*, p. 182.
72. A. Deane-Drummond, *Return Ticket* (London: Collins, 1953), p. 81 and Tony Deane-Drummond letter to the author 18 May 2005.
73. NA, WO 203/2418, Major General Down, 50 Independent Parachute Brigade, 17 October 1944.
74. Dover, *The Sky Generals*, p. 87.
75. Baynes, *Urquhart of Arnhem*, p. 69, Frost, *A Drop Too Many*, pp. 194–195, Deane-Drummond, *Arrows Of Fortune*, p. 88.
76. B.L. Montgomery, *The Memoirs Of Field Marshal Viscount Montgomery Of Alamein* (London: Collins, 1958), p. 76.
77. Baynes, *Urquhart Of Arnhem*, pp. 52—65.
78. ibid, p. 55.
79. Frost, *A Drop Too Many*, p. 194.

NOTES

80. Baynes, *Urquhart Of Arnhem*, p. 72.
81. Urquhart, *Arnhem*, p. 18.
82. Baynes, *Urquhart Of Arnhem*, p. 94.
83. Tony Deane-Drummond letters to the author 18 May and 3 July 2005.
84. Baynes, *Urquhart Of Arnhem*, p. 94. Chatterton thought a *coup de* main landing was possible but was made to feel like 'a bloody murderer' for suggesting it. However, Chatterton makes no reference to this episode in his autobiography. G. Chatterton, *The Wings Of Pegasus* (London: MacDonald, 1962).
85. Only 1 Parachute Brigade and 1 Airlanding Brigade landed on the first day of the operation. 4 Parachute brigade and the Polish Parachute brigade were not due to land until the second and third days of the operation. For a full account of the battle for Arnhem see M. Middlebrook, *Arnhem 1944*, (London: Penguin, 1995).
86. MAF, 1 Airborne Division Report on Operation Market Garden, 10 January 1945, Part IV, Annex N, War Diary of 1 Parachute Brigade.
87. MAF, 16 SS Panzer Grenadier Depot and Reserve Battlion, War Diary 17 September to 7 October 1944.
88. French, *Raising Churchill's Army*, p. 194.
89. Urquhart, *Arnhem*, p. 56.
90. Tony Deane-Drummond letter to the author 18 May 2005.
91. For an abridged and non-critical account of Browning's life see Dover, *The Sky Generals*, pp. 38–56.
92. Otway, *Airborne Forces*, p. 39.
93. Chatterton, *The Wings of Pegasus*, p. 19.
94. Gale, *Call To Arms*, p. 120.
95. IWM 99/18/1, Papers of Major Pope, September 1939 to February 1944.
96. Bond, *Chief of Staff*, p. 193.
97. Horrocks, *Corps Commander*, p. 109.
98. J.M. Gavin, *On to Berlin* (New York: Bantam, 1979), pp. 91 & 187.
99. IWM, 99/18/1, Papers of Major Pope: F.A.M. Browning, *Discipline – The Only Road to Victory*, 1942.
100. IWM, Papers of Brigadier P.N.R. Stewart-Richardson MBE.
101. One photograph, often reproduced, sums up his apparent obsession with discipline and appearance. On 19 May 1944 King George VI, Queen Elizabeth and Princess Elizabeth visited 6 Airborne Division accompanied by Browning. In the photograph a line of soldiers from 224 Parachute Field Ambulance stand to attention, ready for battle, with their equipment laid out to view. The King and Queen stands in front of the men, smiling, relaxed and apparently interested in what the men have to say. Meanwhile Browning is behind the men, bent over and using his swagger stick to inspect the contents of their backpacks. P. Harclerode, *Go To It! The Illustrated History of 6th Airborne Division* (London: Caxton, 1990), p. 50 or Bernage, *Red Devils In Normandy*, p. 14.
102. Frost, *Nearly There: The Memoirs Of John Frost Of Arnhem Bridge*, p. 71.
103. IWM, Misc 35, Item 638, Address by Major General F.A.M Browning to the Officers of the Airborne Division, 1942.
104. NA, AIR 39/39, Browning to Barratt, 7 May 1942.
105. Otway, *Airborne Forces*, pp. 341–343.
106. NA, WO 32/9778, Organisation and Duties of Airborne Organisation, 19 October 1942 and Browning to Crawford (Director Air), 26 March 1943.

107. IWM 99/18/1, Papers of Major Pope, September 1939 to February 1944.
108. NA, WO 204/4585, Correspondence between Eisenhower and Clark concerning planning for Operation TORCH, 8 September–9 October 1942.
109. NA, DEFE 2/542, Operation Rutter planning diary, 21 April 1942 & 13 May 1942.
110. NA, WO 32/9778, Browning to GHQ Home Forces, Future Organisation, Airborne Forces, 6 August 1942.
111. NA, WO 106/4176, Mountbatten to Ismay, Operation 'Blazing', 5 May 1942.
112. NA, WO 204/1396, COS Allied Force HQ, Employment of Airborne Forces in CORKSCREW, 17 May 1943.
113. NA, WO 106/3877, Report on Airborne Operations 'HUSKY', 24 July 1943 and WO 106/4133, Operation BITING, 3 March 1942. Browning's report on Operation BITING was largely uncritical although post HUSKY he highlighted the lack of training and the requirement for a dedicated RAF support organisation.
114. The only other candidate was the commander of XVII US Airborne Corps, Lieutenant General Matthew Ridgeway but his HQ had already been fighting in Normandy, controlling 82 and 101 Airborne Divisions. Gavin, *On to Berlin*, pp. 90–92.
115. Middlebrook, *Arnhem 1944*, p.12.
116. Horrocks insisted that Browning and he 'took all the major decisions together without any semblance of friction.' NA, CAB 106/1054, The Battle of Arnhem – Notes Made By Lt Gen Sir B. Horrocks, undated, p. 5. However Browning's orders to 82nd US Airborne Division for Operation MARKET gave the capture of the high ground at Groesbeek, location for the Corps HQ, equal prominence to the capture of Nijmegen Bridge. These orders almost certainly delayed the capture of the bridge and therefore slowed XXX Corps advance towards Arnhem. NA, WO 176/366, Op MARKET Operation Instruction No. 1, 13 September 1944.
117. NA, WO 176/366, Op MARKET Operation Instruction No. 1, 13 September 1944.
118. ABFM, File 63, Down to Gale, 10 January 1945.
119. LHCMA, 6/2/56, Browning to CIGS, 25 December 1945.
120. Otway, *Airborne Forces*, pp. 329 & 351. Another legacy of Browning was the deployment of I British Airborne Corps HQ on Exercise BUZZ in order to test its procedures in December 1944. NA, WO 219/4981, Airborne Corps Exercise "BUZZ", 9 December 1944.
121. Frost, *Nearly There*, p. 71.
122. R. Gale, *With the 6th Airborne Division In Normandy*, (London: Samson Low, 1948), p. 27.
123. Horrocks, *Corps Commander*, p. 22.
124. Sixsmith, *British Generalship*, p. 171.
125. D. Fraser, *And We Shall Shock Them* (London: Hodder & Stoughton, 1983), p. 103.
126. Montgomery, *The Memoirs Of Field Marshal Viscount Montgomery Of Alamein*, p. 77.
127. IWM, Papers of Lieutenant General Sir Napier Crookenden KCB DSO OBE DL and 99/18/1, Papers of Major A.A.K. Pope.

Notes

128. Otway, *Airborne Forces*, pp. 34–39.
129. NA, AIR 39/39, Formation of HQ, Airborne Division, 2 December 1941.
130. Otway, *Airborne Forces*, p. 40.
131. NA, WO 32/9778, Organisation and Duties of Airborne Forces and Airborne Division, 19 November 1942.
132. NA, WO 32/9778, Organisation, Airborne Forces, 19 October 1942.
133. Otway, *Airborne Forces*, pp. 100 & 111.
134. Chatterton, *Wings of Pegasus*, p. 38.
135. Otway, *Airborne Forces*, pp. 137–139.
136. NA, WO 219/4981, Exercise "BUZZ" Instructions, 14 December 1944.
137. Gavin, *On to Berlin*, pp. 90–91.
138. NA, AIR 32/2, Staff Duties in Connection With the Use of Parachute Troops, 1 March 1941.
139. ABFM, 1/15, COS(44) 230(O), Policy as to the Organization and Employment of Airborne Troops, 6 March 1944.
140. Otway, *Airborne Forces*, pp. 74, 81 & 89.
141. ibid, p. 118.
142. NA, WO 204/1818, Organisation of Airborne Forces Staff, 20 July 1943.
143. NA, WO 204/1818, Report on Airborne Operations 'HUSKY', 24 July 1943.
144. NA, WO 204/1818, Employment of Airborne Forces, 2 August 1943.
145. ABFM, 1/15, COS(43) 552(O), Report on the Employment of Airborne Forces, 20 September 1943.
146. ABFM, 1/15, COS(44) 230(O), Policy as to the Organization and Employment of Airborne Troops, 6 March 1944.
147. Otway, *Airborne Forces*, pp. 201–205.
148. French, *Raising Churchill's Army*, p. 164.
149. J. Keegan, *The Mask of Command*, (London: Penguin, 1988), p. 336.
150. Otway, *Airborne Forces*, pp. 262 & 297.
151. Sixsmith, *British Generalship*, p. 296.
152. B.H. Liddell Hart, *Thoughts on War* (London: Faber & Faber, 1944), p. 226.
153. French, *Raising Churchill's Army*, p. 283.
154. LHCMA, letter Gale to Liddell Hart, 16 April 1952.
155. NA, WO 205/751, Formation of 6 Airborne Division, 6 August 1943.
156. Figures extracted from Otway, *Airborne Forces*, pp. 191, 283 & 319.
157. Holmes, *Firing Line*, p. 349.
158. ABFM, File 212, letter from Hackett published in The Times, 25 June 1977 and unpublished letter Taylor to The Times, 7 July 1977.

Chapter Six
1. H.R. Winton, 'On Military Change' in H.R. Winton and D.R. Mets, eds., *The Challenge Of Change: Military Institutions and New Realities, 1918–1941* (London: University of Nebraska Press, 2000) p. xii.
2. The various debates surrounding the concept of mechanised warfare and the purpose of the tank are described in S.P. Rosen, *Winning the Next War* (New York: Cornell University Press, 1991), pp. 109–129, H.R. Winton, 'Tanks, Votes and Budgets' in Winton and Mets, eds., *The Challenge Of Change*, pp. 74–107 and W. Murray, 'Armoured Warfare' in W. Murray and A.R. Millett, eds., *Military Innovation In The Interwar Period* (New York: Cambridge University Press, 1996), pp. 6–49.

3. J.P. Harris, 'Obstacles to Innovation and Readiness' in Murray and Hart Sinnreich, eds., *The Past As Prologue*, p. 214.
4. See for example D.E. Showalter, 'Military Innovation and the Whig Perspective of History' in Winton and Mets, eds., *The Challenge Of Change*, p. 225 and J.P. Harris 'Obstacles to Innovation and Readiness' in W. Murray and R. Hart Sinnreich, eds., *The Past As Prologue*, p. 215.
5. R.A. Mason, 'Innovation and the Military Mind', *Air University Review*, Vol XXXVII, No. 2, January – February 1986, p. 39.
6. Rosen, *Winning the Next War*, p. 11.
7. IWM, 69/34/1, P R C Groves to Major General Salmond CMG DSO, 17 November 1917, p. 2.
8. IWM, 69/34/1, p. 3.
9. For a description of Soviet airborne development before the war see Buckingham, *Paras*, pp. 38–43 and Glantz, *A History of Soviet Armed Forces*, pp. 1–46.
10. NA, WO 32/4371, MO1 to DDMO, 15 December 1936.
11. NA, WO 32/4371, Parachute Group: Employment, Training and Organisation, 19 April 1938.
12. Liddell Hart advised both Alfred Duff Cooper and Leslie Hore-Belisha as successive Ministers of War through 1937. H.R.Winton, 'Tanks, Votes and Budgets' in H.R. Winton and D.R. Mets, *The Challenge Of Change: Military Institutions and New Realities, 1918–1941* (London: University of Nebraska Press, 2000), pp. 94–96.
13. LHCMA, Liddell: 12/1937/1, Some Important Technological Military Developments, 16 April 1937.
14. NA, CAB 84/19/29, JPS to Churchill, Future Operational Planning, 20 September 1940.
15. NA, CAB 84/19/112, 'Future Plans: Basic Requirements', 26 September 1940.
16. NA, CAB 65/7/8, Conclusions of Meeting of the War Cabinet, 9 May 1940.
17. NA, AIR 2/7470, Actions Required for Providing Airborne Forces, 5 August 1941.
18. ———, *Military Training Pamphlet No.50, Airborne Troops*, August 1941 (NA, WO231/126).
19. ibid, pp.1-5.
20. ———, *Army Training Instruction No.5, Employment of Parachute Troops*, 1941, (LHCMA LH 15/8/148).
21. NA, AIR 23/6110, HQ RAF Middle East Intelligence Report, German Air-Borne Attack on Crete, 1 November 1941.
22. NA, CAB 84/20/44, Future Operational Planning, 5 October 1940. Wilmot states that Churchill initiated the planning process on the 5 October 1940 although clearly considerable work had already been completed prior to this date, C. Wilmot, *The Struggle For Europe* (London: Collins, 1952), p. 97.
23. NA, CAB 84/20/212, 'Future Plans: Basic Requirements', 18 October 1940.
24. NA, CAB 120/262, meeting held at the Air Ministry, 5 September 1940.
25. NA, CAB 120/262, Air Ministry, The Employment of Airborne Forces, 2 September 1940.
26. NA, AIR 32/2, meeting held at the Air Ministry, 5 September 1940.

27. NA, WO 32/9778, DMO&P, 6 January 1941 and Air 2/7470, Airborne Troops – Policy For, 10 January 1941.
28. NA, AIR 2/7470, Airborne Troops – Policy For, 14 January 1941.
29. ABFM, File 1A/2, Summary of Service of Lt Col J.F.Rock, undated.
30. NA, WO 32/9778, Rock to War Office SD4, 15 January 1941.
31. NA, AIR 32/2, meeting held at the Air Ministry, 5 September 1940.
32. NA, AIR 2/7470, DMC to VCIGS, 5 February 1941.
33. NA, AIR 2/7470, DMC to SD4, 17 February 1941.
34. NA, AIR 2/7470, Policy for the Provision of Airborne Forces, 24 March 1941.
35. NA, AIR 23/6110, HQ RAF Middle East Intelligence Report, German Air-Borne Attack on Crete, 1 November 1941.
36. Otway, *Airborne Forces*, pp. 11–12. For a full account of the German invasion of Crete see C. MacDonald, *The Lost Battle: Crete 1941*, (London: Pan, 2002) and A. Beevor, *Crete: The Battle and the Resistance* (London: Penguin, 1991).
37. A.E. Sítek, and V. Blunt, *The Flying Soldier: The Air Requirements Of Airborne Forces* (London: Alliance Press, 1944), pp. 13–14.
38. Gale, *Call To Arms*, p. 133.
39. NA, CAB 120/262, Cherwell to Churchill, 29 April 1942.
40. ———, *Airborne Operations No.1 – General*, 1943 (ABFM).
41. NA, AIR 20/2437, CIGS, The Value of Airborne Forces, 10 October 1942.
42. In this thesis the term 'deep' is used to describe operations against enemy forces and installations not immediately engaged in the close battle. The targets may not even be military, such as the Tregino aqueduct in Italy. This is a different concept of employment to the Soviet model of deep battle and the airborne arm's place within it. In the Soviet concept deep airborne operations involved paratroops landing in the immediate rear of an enemy's front line in coordination with a frontal assault. R. Simpkin, *Deep Battle: The Brainchild of Marshal Tukhachevskii* (Brasseys: London, 1987) pp. 42–43.
43. NA, AIR 32/2, CLE, Brief Appreciation of the Envisaged Functions of Airborne Forces: Theory, 31 October 1940.
44. Churchill liked the phrase 'spearhead' and had circled and underlined it throughout a previous document; NA, CAB 120/262, minutes of meeting held at the Air Ministry, 5 September 1940.
45. NA, DEFE 2/791, agenda for meeting held at the Air Ministry, 11 December 1940.
46. Liddell Hart, *Thoughts on War*, p. 247.
47. LHCMA, Liddell: 12/1941/5, Possibilities and Problems of Invasion, 19 November 1941.
48. NA, CAB 120/262, Amery to Churchill, 6 October 1941.
49. Miksche, *Paratroops*, pp. 56–68.
50. ibid, pp. 62 and 64.
51. In 1938 Liddell Hart was appointed by Churchill as adviser to the 'Focus' group, a small group of eminent men brought together to discuss the growing crisis in Europe. There is no evidence that Liddell Hart contributed any thoughts connected with airborne forces during that period. Lewin, *Churchill as Warlord*, p.15. Miksche did in fact become a person of interest to the central military establishment after the publication of his book but only because one

of his illustrations (ibid, p. 64, illustration 10) showed a remarkable resemblance to the allied plans for OVERLORD including the location. Dank, *The Glider Gang*, p. 103.
52. NA, AIR 2/7470, Air Ministry, Provision of Airborne Troops, 14 February 1941.
53. NA, AIR 2/7470, D Plans to CAS, 24 June 1941.
54. NA, CAB 120/262, Chiefs of Staff Committee, 6 August 1940.
55. Hough, *Mountbatten: Hero Of Our Time*, p. 146.
56. NA, DEFE 2/791, Employment of Airborne Troops, 4 September 1940.
57. NA, DEFE 2/546, Operation 'Rutter' and 'Jubilee' – Notes on Principal Changes in the Military Plan, 14 September 1942.
58. NA, WO 32/9778, Browning to GHQ Home Forces, Future Organisation, Airborne Forces, 6 August 1942.
59. NA, AIR 32/2, draft paper by Rock, Training and Organisation of Airlanding Troops, July 1940.
60. Frost, *Nearly There*, p. 71.
61. Frost, *A Drop Too Many*, p. 84.
62. Otway, *Airborne Forces*, p. 80.
63. Slessor maintained his scepticism towards airborne forces throughout the war and beyond, publicly questioning their effectiveness in 1948. Slessor, 'Some Reflections on Airborne Forces', pp. 161–166.
64. For a full account of Operation FRESHMAN see R. Wiggan, *Operation Freshman* (London: Kimber, 1986).
65. NA, AIR 5/278, Minute from CAS to Secretary of State for Air, 22 August 1922.
66. NA, AIR 5/268, Report on the Development of Soaring Flight in Germany, 18 September 1922.
67. NA, AIR 5/278, Minute from Director of Research to CAS, 6 September 1922 and The Birmingham Post, 18 September 1922.
68. NA, AIR 5/278, Minute from Air Member for Supply and Research to Under Secretary of State for Air, 22 December 1922.
69. Reg Leach, letter to the author, 4 December 2006.
70. NA, WO 199/438, Possible German Use of Gliders Against This Country, 19 June 1940.
71. NA, CAB 120/262, meeting held in the Air Ministry, 7 September 1940.
72. P.W. Thompson, 'How The Germans Took Fort Eben Emael', *United States Army Infantry Journal*, August 1942, pp. 24–26.
73. NA, AIR 20/5256, meeting held in the Air Ministry, 25 July 1940.
74. NA, AIR 20/3378, ACAS(T), 4 September 1940.
75. NA, AIR 20/3378, ACAS(T) to ACAS(R), 4 September 1940.
76. NA WO 199/438, Possible German Use of Gliders Against This Country, 19 June 1940 and AIR 20/5256, meeting held at the Air ministry, 25 July 1940 and AIR 2/7470, draft COS paper, Policy for Airborne Forces, 17 March 1941.
77. NA, WO 199/438, Air Ministry to the War Office, 18 June 1940.
78. NA, AIR 37/280, Notes on the Use of Gliders V. Parachutists in Mountainous Country, 18 October 1942.
79. NA WO 199/438, Air Intelligence note, Possible German Use of Gliders Against This Country, 19 June 1940.
80. NA, AIR 20/5256, The Military Employment of Gliders, 25 July 1940.

NOTES

81. NA, AIR 39/52, The Operational use of Gliders, 15 November 1942.
82. NA, AIR 39/52, meeting, The Operational Use of Gliders, 27 November 1941.
83. NA, AIR 2/7551, AFEE paper, Horsa landing, 4 April 1942.
84. NA, AIR 2/7470, Goddard to Nye, 7 March 1941.
85. NA, CAB 120/262, Present Situation in Respect of the Development of Parachute Training, 12 August 1940.
86. NA, Air 32/2, Characteristics of Airborne Forces, November 1940.
87. NA, DEFE 2/791, Air-Borne Forces, 11 December 1940.
88. ———, *Airborne Operations, A German Appraisal* (Washington: United States Army Department, 1951), p. 11.
89. NA, CAB 120/262, meeting held in the Air Ministry, 7 September 1940.
90. Otway, *Airborne Forces*, p. 7.
91. NA, AIR 23/6110, The German Air-Borne Attack on Crete, 1 November 1941.
92. NA, AIR 32/2, meeting held at the Air Ministry, 5 September 1940.
93. NA, CAB 120/262, meeting held in the Air Ministry, 7 September 1940 and DEFE 2/791, Notes by the General Staff on the Employment of Airborne Troops, 4 September 1940.
94. NA, WO 205/751, Formation of 6 Airborne Division, 6 August 1943. This figure is based on fighting battalions only within a division: six battalions of paratroops totalling 3192 men against three airlanding battalions totalling 1634 men.
95. No glider and airlanding doctrine was published until Airborne Operations Pamphlet No.1 in May 1943, which was too late to have any influence over the planning for HUSKY.
96. Figures extracted from D.H. Wood, *A Noble Pair of Brothers* (———:———, 1996), p. 43 and C. Smith, *The History Of The Glider Pilot Regiment* (London: Leo Cooper, 1992), pp. 55–66.
97. Otway, *Airborne Forces*, pp. 123 & 130.
98. NA, WO 204/4220, Report on the Proceedings of a Board of Officers convened on 23 July 1943 and ABFM, 1/15, COS(43) 552(O), Report on the Employment of Airborne Forces, 20 September 1943.
99. NA, WO 205/26, Report on 6 Airborne Division During Operation NEPTUNE, 14 November 1944.
100. NA, WO 205/751, based on war fighting establishments within 6 Airborne Division.
101. Harrison Place, Military Training In The British Army 1940–1944, pp. 150–152.
102. Gliders were also attached to 9 Parachute Battalion for the assault on the Merville but these were manned by paratroops of the battalion rather than by airlanding troops. The reasons for this decision are unclear as there were more airlanding troops available for D-Day from 6th Airborne Division than were used. The creation of the Armoured Reconnaissance Group is a good example of battle-grouping within 6th Airborne Division on D-Day but was constituted from the Airborne Armoured Reconnaissance Regiment with a company from 12 Battalion, The Devonshire Regiment with supporting arms and did not represent a mix of parachute and airlanding infantry.
103. An airlanding brigade has twelve companies but during Operation MARKET two companies of 2 Battalion, The South Staffordshire Regiment were not landed until the second day due to a lack of gliders.

104. MAF, 29/E/02, 1 Airborne Division Report on Operation Market Garden, Part III, 10 January 1945.
105. NA, WO 205/947, 6 Airborne Division Report on Operation VARSITY, undated, 1945.
106. ———, *Army Doctrine Publication, 'Land Operations'*, p. 156.
107. Otway, *Airborne Forces*, pp. 413–420.
108. Gale, *Call To Arms*, p. 154.

Chapter Seven
1. See for example B.R. Posen, *The Sources of Military Doctrine* (London: Cornell, 1984), pp. 55–182.
2. Harrison Place, *Military Training In The British Army 1940–1944*, p. 168.
3. ———, *Army/Air Operations Pamphlet No.4 – Airborne/Air Transported Operations*, 1945 (Airborne Forces Museum File).
4. French, *Raising Churchill's Army*, pp. 201–202 and Harrison Place, *Military Training In The British Army 1940–1944*, p. 16.
5. Posen, *The Sources of Military Doctrine*, pp. 222–228.
6. Rosen, *Winning the Next War*, p. 11.
7. Quoted in W. Murray, 'Armoured Warfare' in Murray and Millett, eds., *Military Innovation in the Interwar Period*, p. 25.
8. Buckingham, *The Establishment and Initial Development of British Airborne Forces*, p. 300.
9. Judkins, *Making Vision into Power*, p. 562.
10. Hart, *Clash of Arms*, p. 412.

Index

Adam, Ronald, 98
Airborne Forces, Foreign
  France, 93
  Germany, 22, 31, 34, 54, 80, 93, 176–179, 190, 192
  India, 27, 46–50, 56
  Soviet Union, 15–16, 174
  United States of America, 65, 112, 116
    XVIII US Airborne Corps, 168
    82nd US Airborne Division, 18, 139, 156
Airborne Forces Establishment (AFE), 26, 38, 88, 162
Airborne Forces Experimental Establishment (AFEE), 54, 58, 64, 70, 71, 72, 74
Aircraft
  Albermarle, Armstrong-Whitworth, 45, 71, 77, 121
  Beaufighter, Bristol, 77
  Bombay, Bristol, 66
  C-82, Fairchild, 91
  C-93, Budd (Conestoga), 91
  Dakota, Douglas C-47, 68, 73, 75, 76, 77–78, 80, 91, 105, 115, 121, 135, 189
  DC-3, Douglas, 67, 73
  Flamingo, de Havilland, 66
  Frobisher, de Havilland, 66
  Halifax, Handley-Page, 45, 69, 71, 75, 77, 80, 84, 121
  Harrow, Handley Page, 67
  Hector, Hawker, 26, 120
  Hertfordshire, de Havilland, 45, 66, 67
  Hudson, Lockheed, 73
  Hurricane, Hawker, 44
  JU-52, Junkers, 66, 80
  Lancaster, Avro, 71, 73, 75, 80
  Liberator, Consolidated, 73
  Lysander, Westland, 120
  Manchester, Avro, 45, 68, 69, 71
  Mosquito, de Havilland, 88
  Spitfire, Supermarine, 44, 77
  Stirling, Shorts, 45, 69, 75, 121
  Wellington, Vickers, 44, 70, 72, 73, 75, 120
  Whitley, Armstrong-Whitworth, 24, 26, 44, 66, 68, 70, 73, 76, 77, 105, 115, 185
Air Ministry, 31–40
  Air Member for Research and Development, 41
  Air Member for Supply, 41
  Director General of Equipment (DGE), 42–44
  Director of Military Co-operation, 37, 40, 72, 86
  Director of Operational Requirements (DOR), 69
  Director of Scientific Research (DSR), 81
Alexander, Harold, 30, 77, 89, 138–140, 146, 193
Allied Forces Headquarters (AFHQ), 136, 166, 193
Amery, Leo, 46–50, 151, 183–184, 203
Anderson, Kenneth, 165
Army Operational Research Group (AORG), 55, 198

Attlee, Clement, 23
Auchinleck, Claude, 48, 157

Basic Flying Training School (BFTS), 113
Beaverbrook, Lord, 33, 34, 42–46, 47, 67, 91
Beddel-Smith, Walter, 139
Bevan, Ernest, 41
Bols, Eric, 149, 151
Bourne, Alan, 39
Brereton, Louis, 159, 167, 168
British Airborne Formations and Units
  Corps
    1st Airborne Corps, 149, 158–160, 164, 168, 171
    Army Air Corps (AAC), 130
  Divisions
    1st Airborne Division, 19, 26, 30, 100, 102, 108, 120, 131, 133, 136, 138–139, 144, 145, 149, 150, 151, 153–155, 159–160, 163, 164, 165, 166, 170, 171, 188, 195, 196
    6th Airborne Division, 15, 30, 98, 100, 102, 108, 126, 131, 133, 136, 148, 149, 151, 160, 163, 164, 170, 171, 188, 194, 196–197
    44th Indian Airborne Division, 50, 130, 151, 158, 160
  Brigades
    1 Airlanding Brigade, 77, 101, 118, 133, 142, 150, 153, 161, 193, 194
    1 Parachute Brigade, 26, 57, 59, 64, 70, 72, 96, 101, 125, 130, 133, 142, 148, 149, 152, 153, 156, 161, 163, 165, 171, 187–188, 193, 194, 195
    2 Independent Parachute Brigade, 50, 100, 131, 133, 138–139, 149
    3 Parachute Brigade, 148, 149
    4 Parachute Brigade, 100, 108, 149
    5 Parachute Brigade, 97, 149
    6 Airlanding Brigade, 125, 150, 161, 168
    50 Indian Parachute Brigade, 99, 157
  Battalions/Regiments
    1 Parachute Battalion, 25–26, 97, 147
    2 Parachute Battalion, 58, 153
    3 Parachute Battalion, 153–155
    7 Parachute Battalion, 101, 195
    9 Parachute Battalion, 64, 94, 128, 168
    11 Special Air Service Battalion, 25, 97, 99, 107, 123, 149
    12 Parachute Battalion, 101
    13 Parachute Battalion, 101
    154 Parachute Battalion, 151
    Airborne Armoured Reconnaissance Regiment, 62
    Glider Pilot Regiment (GPR), 109–118, 127
    No.2 Commando, 94–99, 101, 103–105, 109
    Oxfordshire and Buckinghamshire Light Infantry, 2 Battalion, 128, 195
    South Staffordshire Regiment, 2 Battalion, 159
  Companies/Squadrons
    21 Independent Parachute Squadron, 58
British Army Formations and Units (less Airborne)
  Army Group
    Fifteenth Army Group, 139–140, 165, 166
    Twenty-First Army Group, 167
  Army
    First Army, 165, 168, 188
    Second Army, 17, 136, 145, 159, 170
    Eighth Army, 141–142, 152, 165–166
  Corps
    I Corps, 136–137
    III Corps, 164
    V Corps, 141
    VIII Corps, 136–137
    XII Corps, 141, 152
    XIII, 166
    XXX Corps, 136, 159
  Division
    1st Division, 168
    3rd Division, 141, 152
    4th Division, 141
    10th Indian Division, 47
    51st Highland Division, 152

INDEX

Brigade
　1 Special Service (SS) Brigade, 136–137, 170
　24 Guards Brigade, 155
　31 Independent Brigade Group, 101, 133, 150, 194
　128 Infantry Brigade, 155
　231 (Malta) Infantry Brigade, 152
Battalion
　Green Howards, 10 Battalion, 101
　King's Own Royal Regiment, 1 Battalion, 47
　Somerset Light Infantry, 10 Battalion, 101
　South Lancashire Regiment, 2/4 Battalion, 101
Brooke, Alan, 28, 36, 49, 145
Browning, Frederick, 26, 50, 74, 87, 98, 120–121, 124, 140, 150, 151, 155–160, 161, 163–166, 171, 186–187, 193, 203, 204
Burnett-Stuart, John, 202

Central Landing Establishment (CLE), 24, 26, 54, 58, 69, 70, 71, 85, 88, 105, 108, 109, 118–129, 164, 178, 181
Central Landing School (CLS), 23, 54, 66, 70, 71, 108
Chamberlain, Neville, 41
Chatterton, George, 89, 109–110, 118, 122, 126–127, 150, 155
Cherwell, Lord, 87, 89, 179, 192
Chiefs of Staff (COS) Committee, 23–25, 27–29, 39, 40, 47, 48, 68, 72, 89, 116, 140, 166
Clark, Mark, 140–141
Churchill, Winston, 21–31, 32, 33, 34, 36, 40, 42, 44, 45, 47, 49, 50, 51, 68, 73, 74, 77, 87, 88, 92, 93, 115, 120, 173, 175, 179, 181, 186, 188, 197, 199, 202–203
Combined Chiefs of Staff, 137–139, 165, 166
Combined Operations Command (COC), 39, 40
Cripps, Stafford, 87, 89, 192
Crocker, John, 137–138, 145, 170
Crookenden, Napier, 94, 96, 98, 125, 129, 147, 161, 168

De Guingand, Francis, 141
Deane-Drummond, Tony, 128, 147, 150, 151
Dempsey, Miles, 15, 136–138, 145, 170
Dill, John, 16, 21, 36, 48, 93, 123, 185–186
Director of Combined Operations (DCO), 23, 37, 39, 40, 51, 71, 136, 163, 186–187
Down, Ernest, 50, 149, 151, 158, 159, 203, 204

Eisenhower, Dwight, 138–146, 169, 198
Elementary Flying Training School (EFTS), 113–115

Fender, Robert, 113
First Allied Airborne Army, 159, 167–169, 194, 195, 199
Fitch, 153–154
Flavell, Edward, 149, 165, 188
Freeman, Wilfred, 66
Frost, John, 58, 59, 98, 104, 147, 151, 153, 160, 188, 204
Fuller, J.F.C., 134

Gale, Richard, 28, 38, 57–58, 62, 97–98, 104, 125–126, 135, 148–149, 151, 155, 159, 160, 161, 170, 171, 179, 199, 203, 204
Gavin, James, 156, 158
German Army (less Airborne), 168
　Division
　　9th SS Panzer Division, 154
　Battalion
　　16 SS Panzer Grenadier Depot and Reserve Battalion, 154
Gliders, 24, 79–89, 188–197
　Hadrian, WACO CG4A, 84–85, 118
　Hamilcar, General Aircraft Company, 59, 61–62, 78, 79–89, 122
　Hengist, Slingsby, 83–84
　Horsa, Airspeed, 59, 76, 77, 79–89, 118, 120, 135, 159, 191
　Hotspur, General Aircraft Company, 26, 76, 77, 79–89, 115, 120, 191
　Kite, Kirby, 81

Glider Operational Training Unit (GOTU), 114–115
Glider Training School (GTS), 114–115
Goddard, R.V., 38, 178
Gough, Bill, 99
Groves, 173, 180

Hackett, 'Shan', 149
Hardwick Hall, 96, 162
Harris, Arthur, 28, 34, 111
Heavy Glider Conversion Unit, 114
Hill, James, 147–149, 204
Hollis, Leslie, 42
Hope-Thompson, 149
Hopkinson, Frederick, 138, 150, 158, 160, 163, 166, 171
Horrocks, Brian, 97, 130, 136

India
  Office, 25, 41, 46–50
Inter Services Training and Development Centre (ISTDC), 39
Ismay, Hastings, 15, 24, 25, 29, 31, 93

Jefferson, Alan, 129
Joint Planning Committee (JPC), 175
Joint Planning Staff (JPS), 27, 47, 51, 175–179, 188
Joint Staff Mission, 143
Joubert, Phillip, 46

Keyes, Roger, 23, 39, 40, 67, 68
Kindersley, Hugh, 148–150

Lander, John, 58
Lathbury, Gerald, 38, 149, 150–151, 153
Laycock, R.E., 40
Leigh-Mallory, Trafford, 144, 167
Liddell Hart, Basil, 169, 175, 182–184
Linlithgow, Lord, 46–50, 151
Llewllin, J.J., 26, 28
Lovat, Lord, 170
Lyttleton, Oliver, 76

Maitland-Wilson, Henry, 143
Marshall, George, 139, 143
Micksche, Otto, 183–184
Milne, George, 119

Ministry of Aircraft Production (MAP), 26, 29, 41–46, 48, 55, 56, 66, 67, 71, 72, 76, 79, 86–88, 90–91, 198
Ministry of Supply, 41, 90
Moore-Brabazon, John, 86
Montgomery, Bernard, 18, 130, 131, 140–146, 149, 150, 151, 152, 161, 165, 169, 196, 198
Mountbatten, Louis, 40, 156, 159, 203

Newall, Cyril, 36, 42–43
Newnham, Maurice, 123, 204
Nye, Archibald, 38, 157

O'Connor, Richard, 137–138, 145, 170
Operation
  AVALANCHE, 139
  AXEHEAD, 144
  BEGGAR, 84
  BENEFICIARY, 144
  BITING, 18, 40, 127, 186
  BOXER, 144
  COLOSSUS, 18, 40, 120, 128
  COMET, 144–145
  CORKSCREW, 158
  DEADSTICK, 128
  DRACULA, 50, 117, 183
  DRAGOON, 138
  FRESHMAN, 64, 82, 123, 189
  HANDS-UP, 144
  HASTY, 138
  HUSKY, 18, 19, 30, 38, 40, 50, 59, 60, 64, 77, 84, 85, 87, 89, 105, 116, 117, 118, 128, 131, 138–139, 142–143, 150, 165, 166, 167, 188–189, 193, 194, 195, 198, 199
  INFATUATE, 144
  JUBILANT, 117
  JUBILEE, 30, 158, 186
  LINNET, 144–145
  MALLARD, 60, 128
  MARKET GARDEN, 17, 19, 20, 63, 75, 77, 102, 112, 117, 118, 125, 131, 136, 144–145, 151, 153–155, 158–160, 168, 171, 194, 195, 198, 199, 200
  OVERLORD, 19, 56, 77, 87–88, 96, 102, 116, 117, 127, 136, 143, 152, 194

INDEX

249

PLUNDER, 15, 145
ROUNDUP, 29, 38, 138
RUTTER, 30, 158
SLEDGEHAMMER, 29
SWORDHILT, 144
THURSDAY, 85
TRANSFIGURE, 144
TONGA, 128
TORCH, 28, 50, 59, 60, 64, 74, 77, 89, 105, 130, 136, 138, 140–141, 147, 156, 158, 165, 166, 185, 187, 189, 198
VARSITY, 15, 17, 19, 20, 60, 62, 77, 94, 117, 131, 136, 155, 168, 196, 198, 199, 200, 201, 204
VERITABLE, 18
WILD OATS, 144
Otway, Terence, 128

Parachute Training School (PTS)
    Indian, 108
    No.1, 95, 103–109, 123
    No.4, Middle East, 108
Pearson, Alistair, 98, 146
Pine-Coffin, Geoffrey, 101, 127
Poett, Nigel, 97, 149
Pope, A.A.K., 161
Portal, Charles, 28, 29, 36, 48, 49, 74, 76
Pritt, D.N., 115
Pownall, Henry, 42, 48, 156

Renwick, Robert, 26–27, 62, 87–88, 91, 197
Rock, J.F., 83, 101, 177–178, 181, 187, 203
Rommel, Erwin, 48
Royal Air Force (RAF)
    Commands
        Army Co-operation Command (AC Comd), 37, 38, 46, 52, 86, 110, 162
        Bomber Command, 63–65, 120
        Coastal Command, 63–65, 70
        Maintenance Command, 42–44
    Groups
        22 Group, 110, 119
        38 Group, 121, 126, 142, 167, 189
        46 Group, 74, 121, 142, 167, 189
        70 Group, 58, 162
    Wings
        38 Wing, 26, 120–121, 124, 157, 163, 189
    Squadrons
        295 Squadron, 121, 124
        296 Squadron, 26, 120
        297 Squadron, 26, 120
        298 Squadron, 121, 124

Sayers, W.H., 189
Sinclair, Archibald, 26, 28, 36, 42–46
Slessor, Sir John, 17, 19, 21, 37, 39, 74, 103
Slim, William, 130
South East Asia Command (SEAC), 117
Stephenson, J., 38, 123
Strange, Louis, 39, 204
SYMBOL, Conference, 29, 30, 51, 52, 87, 142, 188, 198

Trenchard, Hugh, 32, 189
Tukhachevskii, Marshal, 16

United States Army (less Airborne)
    Seventh Army, 142
United States Army Air Force (USAAF), 74, 75, 77, 89, 118, 142, 167
Urquhart, 'Roy', 50, 151–155, 160, 171, 194, 196

Wallwork, Jim, 127–128
War Office, 31–40
    Director of Air, 28, 149
    Director of Military Operations and Training (DMO&I), 16
    Director of Military Training (DMT), 16
    Director of Staff Duties (DSD), 16, 38
Wavell, Sir Archibald, 15–16, 46–50, 151
Wingate, Orde, 85
Wright, Lawrence, 126
Wright, Maurice, 80, 189